The History of The Telephone

HERBERT N. CASSON

The History of The Telephone

Herbert N. Casson

© 1st World Library – Literary Society, 2004
PO Box 2211
Fairfield, IA 52556
www.1stworldlibrary.org
First Edition

LCCN: 2004091214

Softcover ISBN: 1-59540-652-2
eBook ISBN: 1-59540-752-9

Purchase *"The History of The Telephone"*
as a traditional bound book at:
www.1stWorldLibrary.org/purchase.asp?ISBN=1-59540-652-2

1st World Library Literary Society is a nonprofit organization dedicated to promoting literacy by:

- Creating a free internet library accessible from any computer worldwide.
- Hosting writing competitions and offering book publishing scholarships.

Readers interested in supporting literacy through sponsorship, donations or membership please contact:
literacy@1stworldlibrary.org
Check us out at: www.1stworldlibrary.org

The History of The Telephone
*contributed by the Mahaney Family
in support of
1st World Library Literary Society*

CONTENTS

PREFACE. 7

I   THE BIRTH OF THE TELEPHONE. 9

II THE BUILDING OF THE BUSINESS. 29

III THE HOLDING OF THE BUSINESS. 52

IV THE DEVELOPMENT OF THE ART. 72

V   THE EXPANSION OF THE BUSINESS. 111

VI NOTABLE USERS OF THE TELEPHONE. 130

VII THE TELEPHONE AND NATIONAL EFFICIENCY. 144

VIII THE TELEPHONE IN FOREIGN COUNTRIES. 160

IX THE FUTURE OF THE TELEPHONE. 178

PREFACE

Thirty-five short years, and presto! the newborn art of telephony is fullgrown. Three million telephones are now scattered abroad in foreign countries, and seven millions are massed here, in the land of its birth.

So entirely has the telephone outgrown the ridicule with which, as many people can well remember, it was first received, that it is now in most places taken for granted, as though it were a part of the natural phenomena of this planet. It has so marvellously extended the facilities of conversation - that "art in which a man has all mankind for competitors" - that it is now an indispensable help to whoever would live the convenient life. The disadvantage of being deaf and dumb to all absent persons, which was universal in pre-telephonic days, has now happily been overcome; and I hope that this story of how and by whom it was done will be a welcome addition to American libraries.

It is such a story as the telephone itself might tell, if it could speak with a voice of its own. It is not technical. It is not statistical. It is not exhaustive. It is so brief, in fact, that a second volume could readily be made by describing the careers of telephone leaders whose names I find have been omitted unintentionally from this book - such indispensable men, for instance, as William R. Driver, who has signed more telephone cheques and larger ones than any other man; Geo. S. Hibbard, Henry W. Pope, and W. D. Sargent, three veterans who know telephony in all its phases; George Y. Wallace, the last survivor of the Rocky Mountain pioneers; Jasper N. Keller, of Texas and New England; W. T. Gentry, the central figure of the Southeast, and the following presidents of telephone companies: Bernard E. Sunny, of Chicago; E. B. Field, of Denver; D. Leet Wilson, of Pittsburg; L. G. Richardson, of Indianapolis; Caspar E. Yost, of Omaha; James E. Caldwell, of Nashville; Thomas Sherwin, of Boston; Henry T. Scott, of San Francisco; H. J. Pettengill, of Dallas; Alonzo Burt, of Milwaukee; John Kil-gour, of Cincinnati; and Chas. S. Gleed, of Kansas City.

I am deeply indebted to most of these men for the information which is herewith presented; and also to such pioneers, now dead, as O. E. Madden, the first General Agent; Frank L. Pope, the noted electrical expert; C. H. Haskins, of Milwaukee; George F. Ladd, of San Francisco; and Geo. F. Durant, of St. Louis.

H. N. C.
PINE HILL, N. Y., June 1, 1910.

CHAPTER I

THE BIRTH OF THE TELEPHONE

In that somewhat distant year 1875, when the telegraph and the Atlantic cable were the most wonderful things in the world, a tall young professor of elocution was desperately busy in a noisy machine-shop that stood in one of the narrow streets of Boston, not far from Scollay Square. It was a very hot afternoon in June, but the young professor had forgotten the heat and the grime of the workshop. He was wholly absorbed in the making of a nondescript machine, a sort of crude harmonica with a clock-spring reed, a magnet, and a wire. It was a most absurd toy in appearance. It was unlike any other thing that had ever been made in any country. The young professor had been toiling over it for three years and it had constantly baffled him, until, on this hot afternoon in June, 1875, he heard an almost inaudible sound - a faint TWANG - come from the machine itself.

For an instant he was stunned. He had been expecting just such a sound for several months, but it came so suddenly as to give him the sensation of surprise. His eyes blazed with delight, and he sprang in a passion of eagerness to an adjoining room in which stood a young mechanic who was assisting him.

"Snap that reed again, Watson," cried the apparently irrational young professor. There was one of the odd-looking machines in each room, so it appears, and the two were connected by an electric wire. Watson had snapped the reed on one of the machines and the professor had heard from the other machine exactly the same sound. It was no more than the gentle TWANG of a clock-spring; but it was the first time in the history of the world that a complete sound had been carried along a wire, reproduced perfectly at the other end, and heard by an expert in acoustics.

That twang of the clock-spring was the first tiny cry of the newborn telephone, uttered in the clanging din of a machine-shop and happily heard by a man whose ear had been trained to recognize the strange voice of the little newcomer. There, amidst flying belts and jarring wheels, the baby telephone was born, as feeble and helpless as any other baby, and "with no language but a cry."

The professor-inventor, who had thus rescued the tiny foundling of science, was a young Scottish American. His name, now known as widely as the telephone itself, was Alexander Graham Bell. He was a teacher of acoustics and a student of electricity, possibly the only man in his generation who was able to focus a knowledge of both subjects upon the problem of the telephone. To other men that exceedingly faint sound would have been as inaudible as silence itself; but to Bell it was a thunder-clap. It was a dream come true. It was an impossible thing which had in a flash become so easy that he could scarcely believe it. Here, without the use of a battery, with no more electric current than that made by a couple of magnets, all the waves of a sound had been carried along a wire and changed back to sound at the farther end. It was absurd. It was incredible. It was something which neither wire nor electricity had been known to do before. But it was true.

No discovery has ever been less accidental. It was the last link of a long chain of discoveries. It was the result of a persistent and deliberate search. Already, for half a year or longer, Bell had known the correct theory of the telephone; but he had not realized that the feeble undulatory current generated by a magnet was strong enough for the transmission of speech. He had been taught to undervalue the incredible efficiency of electricity.

Not only was Bell himself a teacher of the laws of speech, so highly skilled that he was an instructor in Boston University. His father, also, his two brothers, his uncle, and his grandfather had taught the laws of speech in the universities of Edinburgh, Dublin, and London. For three generations the Bells had been professors of the science of talking. They had even helped to create that science by several inventions. The first of them, Alexander Bell, had invented a system for the correction of stammering and similar defects of speech. The second, Alexander Melville Bell, was the dean of British elocutionists, a man of creative brain and a most impressive facility of rhetoric. He was the author of a dozen text-books on the art of speaking correctly, and also of a most ingenious sign-language which he called "Visible Speech." Every letter in the alphabet of this language represented a certain action of the lips and tongue; so that a new method was provided for those who wished to learn foreign languages or to speak their own language more correctly. And the third of these speech-improving

Bells, the inventor of the telephone, inherited the peculiar genius of his fathers, both inventive and rhetorical, to such a degree that as a boy he had constructed an artificial skull, from gutta-percha and India rubber, which, when enlivened by a blast of air from a hand-bellows, would actually pronounce several words in an almost human manner.

The third Bell, the only one of this remarkable family who concerns us at this time, was a young man, barely twenty-eight, at the time when his ear caught the first cry of the telephone. But he was already a man of some note on his own account. He had been educated in Edinburgh, the city of his birth, and in London; and had in one way and another picked up a smattering of anatomy, music, electricity, and telegraphy. Until he was sixteen years of age, he had read nothing but novels and poetry and romantic tales of Scottish heroes. Then he left home to become a teacher of elocution in various British schools, and by the time he was of age he had made several slight discoveries as to the nature of vowel-sounds. Shortly afterwards, he met in London two distinguished men, Alexander J. Ellis and Sir Charles Wheatstone, who did far more than they ever knew to forward Bell in the direction of the telephone.

Ellis was the president of the London Philological Society. Also, he was the translator of the famous book on "The Sensations of Tone," written by Helmholtz, who, in the period from 1871 to 1894 made Berlin the world-centre for the study of the physical sciences. So it happened that when Bell ran to Ellis as a young enthusiast and told his experiments, Ellis informed him that Helmholtz had done the same things several years before and done them more completely. He brought

Bell to his house and showed him what Helmholtz had done - how he had kept tuning-forks in vibration by the power of electro-magnets, and blended the tones of several tuning-forks together to produce the complex quality of the human voice.

Now, Helmholtz had not been trying to invent a telephone, nor any sort of message-carrier. His aim was to point out the physical basis of music, and nothing more. But this fact that an electro-magnet would set a tuning-fork humming was new to Bell and very attractive. It appealed at once to him as a student of speech. If a tuning-fork could be made to sing by a magnet or an electrified wire, why would it not be possible to make a musical telegraph - a telegraph with a piano key-board, so that many messages could be sent at once over a single wire? Unknown to Bell, there were several dozen inventors then at work upon this problem, which proved in the end to be very elusive. But it gave him at least a starting-point, and he forthwith commenced his quest of the telephone.

As he was then in England, his first step was naturally to visit Sir Charles Wheatstone, the best known English expert on telegraphy. Sir Charles had earned his title by many inventions. He was a simple-natured scientist, and treated Bell with the utmost kindness. He showed him an ingenious talking-machine that had been made by Baron de Kempelin. At this time Bell was twenty-two and unknown; Wheatstone was sixty-seven and famous. And the personality of the veteran scientist made so vivid a picture upon the mind of the impressionable young Bell that the grand passion of science became henceforth the master-motif of his life.

From this summit of glorious ambition he was thrown,

several months later, into the depths of grief and despondency. The White Plague had come to the home in Edinburgh and taken away his two brothers. More, it had put its mark upon the young inventor himself. Nothing but a change of climate, said his doctor, would put him out of danger. And so, to save his life, he and his father and mother set sail from Glasgow and came to the small Canadian town of Brantford, where for a year he fought down his tendency to consumption, and satisfied his nervous energy by teaching "Visible Speech" to a tribe of Mohawk Indians.

By this time it had become evident, both to his parents and to his friends, that young Graham was destined to become some sort of a creative genius. He was tall and supple, with a pale complexion, large nose, full lips, jet-black eyes, and jet-black hair, brushed high and usually rumpled into a curly tangle. In temperament he was a true scientific Bohemian, with the ideals of a savant and the disposition of an artist. He was wholly a man of enthusiasms, more devoted to ideas than to people; and less likely to master his own thoughts than to be mastered by them. He had no shrewdness, in any commercial sense, and very little knowledge of the small practical details of ordinary living. He was always intense, always absorbed. When he applied his mind to a problem, it became at once an enthralling arena, in which there went whirling a chariot-race of ideas and inventive fancies.

He had been fascinated from boyhood by his father's system of "Visible Speech." He knew it so well that he once astonished a professor of Oriental languages by repeating correctly a sentence of Sanscrit that had been written in "Visible Speech" characters. While he was living in London his most absorbing enthusiasm was the instruction of a class of deaf-mutes, who could be trained to talk, he believed, by means of the "Visible Speech" alphabet. He was so deeply impressed by the progress made by these pupils, and by the pathos of their dumbness, that when he arrived in Canada he was in doubt as to which of these two tasks was the more important - the teaching of deaf-mutes or the invention of a musical telegraph.

At this point, and before Bell had begun to experiment with his telegraph, the scene of the story shifts from Canada to Massachusetts. It appears that his father, while lecturing in Boston, had mentioned Graham's exploits with a class of deaf-mutes; and soon afterward the Boston Board of Education wrote to Graham, offering him five hundred dollars if he would come to Boston and introduce his system of teaching in a school for deaf-mutes that had been opened recently. The young man joyfully agreed, and on the first of April, 1871, crossed the line and became for the remainder of his life an American.

For the next two years his telegraphic work was laid aside, if not forgotten. His success as a teacher of deaf-mutes was sudden and overwhelming. It was the educational sensation of 1871. It won him a professorship in Boston University; and brought so many pupils around him that he ventured to open an ambitious "School of Vocal Physiology," which became at once a profitable enterprise. For a time there seemed to be little hope of his escaping from the burden of this success and becoming an inventor, when, by a most happy coincidence, two of his pupils brought to him exactly the sort of stimulation and practical help that he needed and had not up to this time received.

One of these pupils was a little deaf-mute tot, five years of age, named Georgie Sanders. Bell had agreed to give him a series of private lessons for $350 a year; and as the child lived with his grandmother in the city of Salem, sixteen miles from Boston, it was agreed that Bell should make his home with the Sanders family. Here he not only found the keenest interest and sympathy in his air-castles of invention, but also was given permission to use the cellar of the house as his workshop.

For the next three years this cellar was his favorite retreat. He littered it with tuning-forks, magnets, batteries, coils of wire, tin trumpets, and cigar-boxes. No one outside of the Sanders family was allowed to enter it, as Bell was nervously afraid of having his ideas stolen. He would even go to five or six stores to buy his supplies, for fear that his intentions should be discovered. Almost with the secrecy of a conspirator, he worked alone in this cellar, usually at night, and quite oblivious of the fact that sleep was a necessity to him and to the Sanders family.

"Often in the middle of the night Bell would wake me up," said Thomas Sanders, the father of Georgie. "His black eyes would be blazing with excitement. Leaving me to go down to the cellar, he would rush wildly to the barn and begin to send me signals along his experimental wires. If I noticed any improvement in his machine, he would be delighted. He would leap and whirl around in one of his `war-dances' and then go contentedly to bed. But if the experiment was a failure, he would go back to his workbench and try some different plan."

The second pupil who became a factor - a very

considerable factor - in Bell's career was a fifteen-year-old girl named Mabel Hubbard, who had lost her hearing, and consequently her speech, through an attack of scarlet-fever when a baby. She was a gentle and lovable girl, and Bell, in his ardent and headlong way, lost his heart to her completely; and four years later, he had the happiness of making her his wife. Mabel Hubbard did much to encourage Bell. She followed each step of his progress with the keenest interest. She wrote his letters and copied his patents. She cheered him on when he felt himself beaten. And through her sympathy with Bell and his ambitions, she led her father - a widely known Boston lawyer named Gardiner G. Hubbard - to become Bell's chief spokesman and defender, a true apostle of the telephone.

Hubbard first became aware of Bell's inventive efforts one evening when Bell was visiting at his home in Cambridge. Bell was illustrating some of the mysteries of acoustics by the aid of a piano. "Do you know," he said to Hubbard, "that if I sing the note G close to the strings of the piano, that the G-string will answer me?" "Well, what then?" asked Hubbard. "It is a fact of tremendous importance," replied Bell. "It is an evidence that we may some day have a musical telegraph, which will send as many messages simultaneously over one wire as there are notes on that piano."

Later, Bell ventured to confide to Hubbard his wild dream of sending speech over an electric wire, but Hubbard laughed him to scorn. "Now you are talking nonsense," he said. "Such a thing never could be more than a scientific toy. You had better throw that idea out of your mind and go ahead with your musical telegraph, which if it is successful will make you a millionaire."

But the longer Bell toiled at his musical telegraph, the more he dreamed of replacing the telegraph and its cumbrous sign-language by a new machine that would carry, not dots and dashes, but the human voice. "If I can make a deaf-mute talk," he said, "I can make iron talk." For months he wavered between the two ideas. He had no more than the most hazy conception of what this voice-carrying machine would be like. At first he conceived of having a harp at one end of the wire, and a speaking-trumpet at the other, so that the tones of the voice would be reproduced by the strings of the harp.

Then, in the early Summer of 1874, while he was puzzling over this harp apparatus, the dim outline of a new path suddenly glinted in front of him. He had not been forgetful of "Visible Speech" all this while, but had been making experiments with two remarkable machines - the phonautograph and the manometric capsule, by means of which the vibrations of sound were made plainly visible. If these could be improved, he thought, then the deaf might be taught to speak by SIGHT - by learning an alphabet of vibrations. He mentioned these experiments to a Boston friend, Dr. Clarence J. Blake, and he, being a surgeon and an aurist, naturally said, "Why don't you use a REAL EAR?"

Such an idea never had, and probably never could have, occurred to Bell; but he accepted it with eagerness. Dr. Blake cut an ear from a dead man's head, together with the ear-drum and the associated bones. Bell took this fragment of a skull and arranged it so that a straw touched the ear-drum at one end and a piece of moving smoked glass at the other. Thus, when Bell spoke loudly into the ear, the vibrations of the drum made tiny markings upon the glass.

It was one of the most extraordinary incidents in the whole history of the telephone. To an uninitiated onlooker, nothing could have been more ghastly or absurd. How could any one have interpreted the gruesome joy of this young professor with the pale face and the black eyes, who stood earnestly singing, whispering, and shouting into a dead man's ear? What sort of a wizard must he be, or ghoul, or madman? And in Salem, too, the home of the witchcraft superstition! Certainly it would not have gone well with Bell had he lived two centuries earlier and been caught at such black magic.

What had this dead man's ear to do with the invention of the telephone? Much. Bell noticed how small and thin was the ear-drum, and yet how effectively it could send thrills and vibrations through heavy bones. "If this tiny disc can vibrate a bone," he thought, "then an iron disc might vibrate an iron rod, or at least, an iron wire." In a flash the conception of a membrane telephone was pictured in his mind. He saw in imagination two iron discs, or ear-drums, far apart and connected by an electrified wire, catching the vibrations of sound at one end, and reproducing them at the other. At last he was on the right path, and had a theoretical knowledge of what a speaking telephone ought to be. What remained to be done was to construct such a machine and find out how the electric current could best be brought into harness.

Then, as though Fortune suddenly felt that he was winning this stupendous success too easily, Bell was flung back by an avalanche of troubles. Sanders and Hubbard, who had been paying the cost of his experiments, abruptly announced that they would pay no more unless he confined his attention to the musical telegraph, and stopped wasting his time on ear-toys that never could be of any financial value. What these two men asked could scarcely be denied, as one of them was his best-paying patron and the other was the father of the girl whom he hoped to marry. "If you wish my daughter," said Hubbard, "you must abandon your foolish telephone." Bell's "School of Vocal Physiology," too, from which he had hoped so much, had come to an inglorious end. He had been too much absorbed in his experiments to sustain it. His professorship had been given up, and he had no pupils except Georgie Sanders and Mabel Hubbard. He was poor, much poorer than his associates knew. And his mind was torn and distracted by the contrary calls of science, poverty, business, and affection. Pouring out his sorrows in a letter to his mother, he said: "I am now beginning to realize the cares and anxieties of being an inventor. I have had to put off all pupils and classes, for flesh and blood could not stand much longer such a strain as I have had upon me."

While stumbling through this Slough of Despond, he was called to Washington by his patent lawyer. Not having enough money to pay the cost of such a journey, he borrowed the price of a return ticket from Sanders and arranged to stay with a friend in Washington, to save a hotel bill that he could not afford. At that time Professor Joseph Henry, who knew more of the theory of electrical science than any other American, was the Grand Old Man of Washington; and poor Bell, in his doubt and desperation, resolved to run to him for advice.

Then came a meeting which deserves to be historic. For an entire afternoon the two men worked together over the apparatus that Bell had brought from Boston, just as Henry had worked over the telegraph before Bell was born. Henry was now a veteran of seventy-eight, with only three years remaining to his credit in the bank of Time, while Bell was twenty-eight. There was a long half-century between them; but the youth had discovered a New Fact that the sage, in all his wisdom, had never known.

"You are in possession of the germ of a great invention," said Henry, "and I would advise you to work at it until you have made it complete."

"But," replied Bell, "I have not got the electrical knowledge that is necessary."

"Get it," responded the aged scientist.

"I cannot tell you how much these two words have encouraged me," said Bell afterwards, in describing this interview to his parents. "I live too much in an atmosphere of discouragement for scientific pursuits; and such a chimerical idea as telegraphing VOCAL SOUNDS would indeed seem to most minds scarcely feasible enough to spend time in working over."

By this time Bell had moved his workshop from the cellar in Salem to 109 Court Street, Boston, where he had rented a room from Charles Williams, a manufacturer of electrical supplies. Thomas A. Watson was his assistant, and both Bell and Watson lived nearby, in two cheap little bedrooms. The rent of the workshop and bedrooms, and Watson's wages of nine dollars a week, were being paid by Sanders and Hubbard. Consequently, when Bell returned from Washington, he was compelled by his agreement to devote himself mainly to the musical telegraph, although his heart was now with the telephone. For exactly three months after his interview with Professor Henry, he continued to plod ahead, along both lines, until, on that memorable hot afternoon in June, 1875, the full TWANG of the clock-spring came over the wire, and the telephone was born.

From this moment, Bell was a man of one purpose. He won over Sanders and Hubbard. He converted Watson into an enthusiast. He forgot his musical telegraph, his "Visible Speech," his classes, his poverty. He threw aside a profession in which he was already locally famous. And he grappled with this new mystery of electricity, as Henry had advised him to do, encouraging himself with the fact that Morse, who was only a painter, had mastered his electrical difficulties, and there was no reason why a professor of acoustics should not do as much.

The telephone was now in existence, but it was the youngest and feeblest thing in the nation. It had not yet spoken a word. It had to be taught, developed, and made fit for the service of the irritable business world. All manner of discs had to be tried, some smaller and thinner than a dime and others of steel boiler-plate as heavy as the shield of Achilles. In all the books of electrical science, there was nothing to help Bell and Watson in this journey they were making through an unknown country. They were as chartless as Columbus was in 1492. Neither they nor any one else had acquired any experience in the rearing of a young telephone. No one knew what to do next. There was nothing to know.

For forty weeks - long exasperating weeks - the telephone could do no more than gasp and make strange inarticulate noises. Its educators had not learned how to manage it. Then, on March 10, 1876, IT TALKED. It said distinctly -

"MR. WATSON, COME HERE, I WANT YOU." Watson, who was at the lower end of the wire, in the basement, dropped the receiver and rushed with wild joy up three flights of stairs to tell the glad tidings to Bell. "I can hear you!" he shouted breathlessly. "I can hear the WORDS."

It was not easy, of course, for the weak young telephone to make itself heard in that noisy workshop. No one, not even Bell and Watson, was familiar with its odd little voice. Usually Watson, who had a remarkably keen sense of hearing, did the listening; and Bell, who was a professional elocutionist, did the talking. And day by day the tone of the baby instrument grew clearer - a new note in the orchestra of civilization.

On his twenty-ninth birthday, Bell received his patent, No. 174,465 - "the most valuable single patent ever issued" in any country. He had created something so entirely new that there was no name for it in any of the world's languages. In describing it to the officials of the Patent Office, he was obliged to call it "an improvement in telegraphy," when, in truth, it was nothing of the kind. It was as different from the telegraph as the eloquence of a great orator is from the sign-language of a deaf-mute.

Other inventors had worked from the standpoint of the telegraph; and they never did, and never could, get any better results than signs and symbols. But Bell worked from the standpoint of the human voice. He cross-fertilized the two sciences of acoustics and electricity. His study of "Visible Speech" had trained his mind so that he could mentally SEE the shape of a word as he spoke it. He knew what a spoken word was, and how it acted upon the air, or the ether, that carried its vibrations from the lips to the ear. He was a third-generation specialist in the nature of speech, and he knew that for the transmission of spoken words there must be "a pulsatory action of the electric current which is the exact equivalent of the aerial impulses."

Bell knew just enough about electricity, and not too much. He did not know the possible from the impossible. "Had I known more about electricity, and less about sound," he said, "I would never have invented the telephone." What he had done was so amazing, so foolhardy, that no trained electrician could have thought of it. It was "the very hardihood of invention," and yet it was not in any sense a chance discovery. It was the natural output of a mind that had been led to assemble just the right materials for such a product.

As though the very stars in their courses were working for this young wizard with the talking wire, the Centennial Exposition in Philadelphia opened its doors exactly two months after the telephone had learned to talk. Here was a superb opportunity to let the wide world know what had been done, and fortunately Hubbard was one of the Centennial Commissioners. By his influence a small table was placed in the Department of Education, in a narrow space between a stairway and a wall, and on this table was deposited the first of the telephones.

Bell had no intention of going to the Centennial

himself. He was too poor. Sanders and Hubbard had never done more than pay his room-rent and the expense of his experiments. For his three or four years of inventing he had received nothing as yet - nothing but his patent. In order to live, he had been compelled to reorganize his classes in "Visible Speech," and to pick up the ravelled ends of his neglected profession.

But one Friday afternoon, toward the end of June, his sweetheart, Mabel Hubbard, was taking the train for the Centennial; and he went to the depot to say good-bye. Here Miss Hubbard learned for the first time that Bell was not to go. She coaxed and pleaded, without effect. Then, as the train was starting, leaving Bell on the platform, the affectionate young girl could no longer control her feelings and was overcome by a passion of tears. At this the susceptible Bell, like a true Sir Galahad, dashed after the moving train and sprang aboard, without ticket or baggage, oblivious of his classes and his poverty and of all else except this one maiden's distress. "I never saw a man," said Watson, "so much in love as Bell was."

As it happened, this impromptu trip to the Centennial proved to be one of the most timely acts of his life. On the following Sunday after-noon the judges were to make a special tour of inspection, and Mr. Hubbard, after much trouble, had obtained a promise that they would spend a few minutes examining Bell's telephone. By this time it had been on exhibition for more than six weeks, without attracting the serious attention of anybody.

When Sunday afternoon arrived, Bell was at his little table, nervous, yet confident. But hour after hour went by, and the judges did not arrive. The day was intensely hot, and they had many wonders to examine. There was the first electric light, and the first grain-binder, and the musical telegraph of Elisha Gray, and the marvellous exhibit of printing telegraphs shown by the Western Union Company. By the time they came to Bell's table, through a litter of school-desks and blackboards, the hour was seven o'clock, and every man in the party was hot, tired, and hungry. Several announced their intention of returning to their hotels. One took up a telephone receiver, looked at it blankly, and put it down again. He did not even place it to his ear. Another judge made a slighting remark which raised a laugh at Bell's expense. Then a most marvellous thing happened - such an incident as would make a chapter in "The Arabian Nights Entertainments."

Accompanied by his wife, the Empress Theresa, and by a bevy of courtiers, the Emperor of Brazil, Dom Pedro de Alcantara, walked into the room, advanced with both hands outstretched to the bewildered Bell, and exclaimed: "Professor Bell, I am delighted to see you again." The judges at once forgot the heat and the fatigue and the hunger. Who was this young inventor, with the pale complexion and black eyes, that he should be the friend of Emperors? They did not know, and for the moment even Bell himself had forgotten, that Dom Pedro had once visited Bell's class of deaf-mutes at Boston University. He was especially interested in such humanitarian work, and had recently helped to organize the first Brazilian school for deaf-mutes at Rio de Janeiro. And so, with the tall, blond-bearded Dom Pedro in the centre, the assembled judges, and scientists - there were fully fifty in all - entered with unusual zest into the proceedings of this first telephone exhibition.

A wire had been strung from one end of the room to the other, and while Bell went to the transmitter, Dom Pedro took up the receiver and placed it to his ear. It was a moment of tense expectancy. No one knew clearly what was about to happen, when the Emperor, with a dramatic gesture, raised his head from the receiver and exclaimed with a look of utter amazement: "MY GOD - IT TALKS!"

Next came to the receiver the oldest scientist in the group, the venerable Joseph Henry, whose encouragement to Bell had been so timely. He stopped to listen, and, as one of the bystanders afterwards said, no one could forget the look of awe that came into his face as he heard that iron disc talking with a human voice. "This," said he, "comes nearer to overthrowing the doctrine of the conservation of energy than anything I ever saw."

Then came Sir William Thomson, latterly known as Lord Kelvin. It was fitting that he should be there, for he was the foremost electrical scientist at that time in the world, and had been the engineer of the first Atlantic Cable. He listened and learned what even he had not known before, that a solid metallic body could take up from the air all the countless varieties of vibrations produced by speech, and that these vibrations could be carried along a wire and reproduced exactly by a second metallic body. He nodded his head solemnly as he rose from the receiver. "It DOES speak," he said emphatically. "It is the most wonderful thing I have seen in America."

So, one after another, this notable company of men listened to the voice of the first telephone, and the more they knew of science, the less they were inclined to believe their ears. The wiser they were, the more they wondered. To Henry and Thomson, the masters of electrical magic, this instrument was as surprising as it was to the man in the street. And both were noble enough to admit frankly their astonishment in the reports which they made as judges, when they gave Bell a Certificate of Award. "Mr. Bell has achieved a result of transcendent scientific interest," wrote Sir William Thomson. "I heard it speak distinctly several sentences. . . . I was astonished and delighted. . . . It is the greatest marvel hitherto achieved by the electric telegraph."

Until nearly ten o'clock that night the judges talked and listened by turns at the telephone. Then, next morning, they brought the apparatus to the judges' pavilion, where for the remainder of the summer it was mobbed by judges and scientists. Sir William Thomson and his wife ran back and forth between the two ends of the wire like a pair of delighted children. And thus it happened that the crude little instrument that had been tossed into an out-of-the-way corner became the star of the Centennial. It had been given no more than eighteen words in the official catalogue, and here it was acclaimed as the wonder of wonders. It had been conceived in a cellar and born in a machine-shop; and now, of all the gifts that our young American Republic had received on its one-hundredth birthday, the telephone was honored as the rarest and most welcome of them all.

CHAPTER II

THE BUILDING OF THE BUSINESS

After the telephone had been born in Boston, baptized in the Patent Office, and given a royal reception at the Philadelphia Centennial, it might be supposed that its life thenceforth would be one of peace and pleasantness. But as this is history, and not fancy, there must be set down the very surprising fact that the young newcomer received no welcome and no notice from the great business world. "It is a scientific toy," said the men of trade and commerce. "It is an interesting instrument, of course, for professors of electricity and acoustics; but it can never be a practical necessity. As well might you propose to put a telescope into a steel-mill or to hitch a balloon to a shoe-factory."

Poor Bell, instead of being applauded, was pelted with a hailstorm of ridicule. He was an "impostor," a "ventriloquist," a "crank who says he can talk through a wire." The London Times alluded pompously to the telephone as the latest American humbug, and gave many profound reasons why speech could not be sent over a wire, because of the intermittent nature of the electric current. Almost all electricians - the men who were supposed to know - pronounced the telephone an impossible thing; and those who did not openly declare it to be a hoax, believed that Bell had stumbled upon some freakish use of electricity, which could never be of any practical value.

Even though he came late in the succession of inventors, Bell had to run the gantlet of scoffing and adversity. By the reception that the public gave to his telephone, he learned to sympathize with Howe, whose first sewing-machine was smashed by a Boston mob; with McCormick, whose first reaper was called "a cross between an Astley chariot, a wheelbarrow, and a flying- machine"; with Morse, whom ten Congresses regarded as a nuisance; with Cyrus Field, whose Atlantic Cable was denounced as "a mad freak of stubborn ignorance"; and with Westinghouse, who was called a fool for proposing "to stop a railroad train with wind."

The very idea of talking at a piece of sheet-iron was so new and extraordinary that the normal mind repulsed it. Alike to the laborer and the scientist, it was incomprehensible. It was too freakish, too bizarre, to be used outside of the laboratory and the museum. No one, literally, could understand how it worked; and the only man who offered a clear solution of the mystery was a Boston mechanic, who maintained that there was "a hole through the middle of the wire."

People who talked for the first time into a telephone box had a sort of stage fright. They felt foolish. To do so seemed an absurd performance, especially when they had to shout at the top of their voices. Plainly, whatever of convenience there might be in this new contrivance was far outweighed by the loss of personal dignity; and very few men had sufficient imagination to picture the telephone as a part of the machinery of their daily work. The banker said it might do well enough for grocers, but that it would never be of any value to banking; and the grocer said it might do well enough for bankers, but that it would never be of any value to grocers.

As Bell had worked out his invention in Salem, one editor displayed the headline, "Salem Witchcraft." The New York Herald said: "The effect is weird and almost supernatural." The Providence Press said: "It is hard to resist the notion that the powers of darkness are somehow in league with it." And The Boston Times said, in an editorial of bantering ridicule: "A fellow can now court his girl in China as well as in East Boston; but the most serious aspect of this invention is the awful and irresponsible power it will give to the average mother-in-law, who will be able to send her voice around the habitable globe."

There were hundreds of shrewd capitalists in American cities in 1876, looking with sharp eyes in all directions for business chances; but not one of them came to Bell with an offer to buy his patent. Not one came running for a State contract. And neither did any legislature, or city council, come forward to the task of giving the people a cheap and efficient telephone service. As for Bell himself, he was not a man of affairs. In all practical business matters, he was as incompetent as a Byron or a Shelley. He had done his part, and it now remained for men of different abilities to take up his telephone and adapt it to the uses and conditions of the business world.

The first man to undertake this work was Gardiner G. Hubbard, who became soon afterwards the father-in-law of Bell. He, too, was a man of enthusiasm rather than of efficiency. He was not a man of wealth or business experience, but he was admirably suited to introduce the telephone to a hostile public. His father had been a judge of the Massachusetts Supreme Court; and he himself was a lawyer whose practice had been mainly in matters of legislation. He was, in 1876, a man of venerable appearance, with white hair, worn long, and a patriarchal beard. He was a familiar figure in Washington, and well known among the public men of his day. A versatile and entertaining companion, by turns prosperous and impecunious, and an optimist always, Gardiner Hubbard became a really indispensable factor as the first advance agent of the telephone business.

No other citizen had done more for the city of Cambridge than Hubbard. It was he who secured gas for Cambridge in 1853, and pure water, and a street-railway to Boston. He had gone through the South in 1860 in the patriotic hope that he might avert the impending Civil War. He had induced the legislature to establish the first public school for deaf-mutes, the school that drew Bell to Boston in 1871. And he had been for years a most restless agitator for improvements in telegraphy and the post office. So, as a promoter of schemes for the public good, Hubbard was by no means a novice. His first step toward capturing the attention of an indifferent nation was to beat the big drum of publicity. He saw that this new idea of telephoning must be made familiar to the public mind. He talked telephone by day and by night. Whenever he travelled, he carried a pair of the magical instruments in his valise, and gave demonstrations on trains and in hotels. He buttonholed every influential man who crossed his path. He was a veritable "Ancient Mariner" of the telephone. No possible listener was allowed to escape.

Further to promote this campaign of publicity, Hubbard encouraged Bell and Watson to perform a series of sensational feats with the telephone. A telegraph wire between New York and Boston was borrowed for half an hour, and in the presence of Sir William Thomson, Bell sent a tune over the two-hundred-and-fifty-mile line. "Can you hear?" he asked the operator at the New York end. "Elegantly," responded the operator. "What tune?" asked Bell. "Yankee Doodle," came the answer. Shortly afterwards, while Bell was visiting at his father's house in Canada, he bought up all the stove-pipe wire in the town, and tacked it to a rail fence between the house and a telegraph office. Then he went to a village eight miles distant and sent scraps of songs and Shakespearean quotations over the wire.

There was still a large percentage of people who denied that spoken words could be transmitted by a wire. When Watson talked to Bell at public demonstrations, there were newspaper editors who referred sceptically to "the supposititious Watson." So, to silence these doubters, Bell and Watson planned a most severe test of the telephone. They borrowed the telegraph line between Boston and the Cambridge Observatory, and attached a telephone to each end. Then they maintained, for three hours or longer, the FIRST SUSTAINED conversation by telephone, each one taking careful notes of what he said and of what he heard. These notes were published in parallel columns in The Boston Advertiser, October 19, 1876, and proved beyond question that the telephone was now a practical success.

After this, one event crowded quickly on the heels of another. A series of ten lectures was arranged for Bell, at a hundred dollars a lecture, which was the first money payment he had received for his invention. His opening night was in Salem, before an audience of five hundred people, and with Mrs. Sanders, the motherly old lady who had sheltered Bell in the days of his experiment, sitting proudly in one of the front seats. A pole was set up at the front of the hall, supporting the end of a telegraph wire that ran from Salem to Boston. And Watson, who became the first public talker by telephone, sent messages from Boston to various members of the audience. An account of this lecture was sent by telephone to The Boston Globe, which announced the next morning -

"This special despatch of the Globe has been transmitted by telephone in the presence of twenty people, who have thus been witnesses to a feat never before attempted - the sending of news over the space of sixteen miles by the human voice."

This Globe despatch awoke the newspaper editors with an unexpected jolt. For the first time they began to notice that there was a new word in the language, and a new idea in the scientific world. No newspaper had made any mention whatever of the telephone for seventy-five days after Bell received his patent. Not one of the swarm of reporters who thronged the Philadelphia Centennial had regarded the telephone as a matter of any public interest. But when a column of news was sent by telephone to The Boston Globe, the whole newspaper world was agog with excitement. A thousand pens wrote the name of Bell. Requests to repeat his lecture came to Bell from Cyrus W. Field, the veteran of the Atlantic Cable, from the poet Longfellow, and from many others.

As he was by profession an elocutionist, Bell was able to make the most of these opportunities. His lectures became popular entertainments. They were given in the largest halls. At one lecture two Japanese gentlemen were induced to talk to one another in their own language, via the telephone. At a second lecture a band played "The Star-Spangled Banner," in Boston, and was heard by an audience of two thousand people in Providence. At a third, Signor Ferranti, who was in Providence, sang a selection from "The Marriage of Figaro" to an audience in Boston. At a fourth, an exhortation from Moody and a song from Sankey came over the vibrating wire. And at a fifth, in New Haven, Bell stood sixteen Yale professors in line, hand in hand, and talked through their bodies - a feat which was then, and is to-day, almost too wonderful to believe.

Very slowly these lectures, and the tireless activity of Hubbard, pushed back the ridicule and the incredulity; and in the merry month of May, 1877, a man named Emery drifted into Hubbard's office from the near-by city of Charlestown, and leased two telephones for twenty actual dollars - the first money ever paid for a telephone. This was the first feeble sign that such a novelty as the telephone business could be established; and no money ever looked handsomer than this twenty dollars did to Bell, Sanders, Hubbard, and Watson. It was the tiny first-fruit of fortune.

Greatly encouraged, they prepared a little circular which was the first advertisement of the telephone business. It is an oddly simple little document to-day, but to the 1877 brain it was startling. It modestly claimed that a telephone was superior to a telegraph for three reasons:

"(1) No skilled operator is required, but direct communication may be had by speech without the intervention of a third person.

"(2) The communication is much more rapid, the average number of words transmitted in a minute, by the Morse sounder being from fifteen to twenty, by telephone from one to two hundred.

"(3) No expense is required, either for its operation or repair. It needs no battery and has no complicated machinery. It is unsurpassed for economy and simplicity."

The only telephone line in the world at this time was between the Williams' workshop in Boston and the home of Mr. Williams in Somerville. But in May, 1877, a young man named E. T. Holmes, who was running a burglar-alarm business in Boston, proposed that a few telephones be linked to his wires. He was a friend and customer of Williams, and suggested this plan half in jest and half in earnest. Hubbard was quick to seize this opportunity, and at once lent Holmes a dozen telephones. Without asking permission, Holmes went into six banks and nailed up a telephone in each. Five bankers made no protest, but the sixth indignantly ordered "that playtoy" to be taken out. The other five telephones could be connected by a switch in Holmes's office, and thus was born the first tiny and crude Telephone Exchange. Here it ran for several weeks as a telephone system by day and a burglar-alarm by night. No money was paid by the bankers. The service was given to them as an exhibition and an advertisement. The little shelf with its five telephones was no more like the marvellous exchanges of to-day than a canoe is like a Cunarder, but it was unquestionably the first place where several telephone wires came together and could be united.

Soon afterwards, Holmes took his telephones out of the banks, and started a real telephone business among the express companies of Boston. But by this time several exchanges had been opened for ordinary business, in New Haven, Bridgeport, New York, and Philadelphia. Also, a man from Michigan had arrived, with the hardihood to ask for a State agency - George W. Balch, of Detroit. He was so welcome that Hubbard joyfully gave him everything he asked - a perpetual right to the whole State of Michigan. Balch was not required to pay a cent in advance, except his railway fare, and before he was many years older he had sold his lease for a handsome fortune of a quarter of a million dollars, honestly earned by his initiative and enterprise.

By August, when Bell's patent was sixteen months old, there were 778 telephones in use. This looked like success to the optimistic Hubbard. He decided that the time had come to organize the business, so he created a simple agreement which he called the "Bell Telephone Association." This agreement gave Bell, Hubbard and Sanders a three-tenths interest apiece in the patents, and Watson one-tenth. THERE WAS NO CAPITAL. There was none to be had. The four men had at this time an absolute monopoly of the telephone business; and everybody else was quite willing that they should have it.

The only man who had money and dared to stake it on the future of the telephone was Thomas Sanders, and he did this not mainly for business reasons. Both he and Hubbard were attached to Bell primarily by sentiment, as Bell had removed the blight of dumbness from Sanders's little son, and was soon to marry Hubbard's daughter.

Also, Sanders had no expectation, at first, that so much money would be needed. He was not rich. His entire business, which was that of cutting out soles for shoe manufacturers, was not at any time worth more than thirty-five thousand dollars. Yet, from 1874 to 1878, he had advanced nine-tenths of the money that was spent on the telephone. He had paid Bell's room-rent, and Watson's wages, and Williams's expenses, and the cost of the exhibit at the Centennial. The first five thousand telephones, and more, were made with his money. And so many long, expensive months dragged by before any relief came to Sanders, that he was compelled, much against his will and his business judgment, to stretch his credit within an inch of the breaking-point to help Bell and the telephone. Desperately he signed note after note until he faced a total of one hundred and ten thousand dollars. If the new "scientific toy" succeeded, which he often doubted, he would be the richest citizen in Haverhill; and if it failed, which he sorely feared, he would be a bankrupt.

A disheartening series of rebuffs slowly forced the truth in upon Sanders's mind that the business world refused to accept the telephone as an article of commerce. It was a toy, a plaything, a scientific wonder, but not a necessity to be bought and used for ordinary purposes by ordinary people. Capitalists treated it exactly as they treated the Atlantic Cable project when Cyrus Field visited Boston in 1862. They admired and marvelled; but not a man subscribed a dollar. Also, Sanders very soon learned that it was a most unpropitious time for the setting afloat of a new enterprise. It was a period of turmoil and suspicion. What with the Jay Cooke failure, the Hayes-Tilden deadlock, and the bursting of a hundred railroad bubbles, there was very little in the news of the day to encourage investors.

It was impossible for Sanders, or Bell, or Hubbard, to prepare any definite plan. No matter what the plan might have been, they had no money to put it through. They believed that they had something new and marvellous, which some one, somewhere, would be willing to buy. Until this good genie should arrive, they could do no more than flounder ahead, and take whatever business was the nearest and the cheapest. So while Bell, in eloquent rhapsodies, painted word-pictures of a universal telephone service to applauding audiences, Sanders and Hubbard were leasing telephones two by two, to business men who previously had been using the private lines of the Western Union Telegraph Company.

This great corporation was at the time their natural and inevitable enemy. It had swallowed most of its competitors, and was reaching out to monopolize all methods of communication by wire. The rosiest hope that shone in front of Sanders and Hubbard was that the Western Union might conclude to buy the Bell patents, just as it had already bought many others. In one moment of discouragement they had offered the telephone to President Orton, of the Western Union, for $100,000; and Orton had refused it. "What use," he asked pleasantly, "could this company make of an electrical toy?"

But besides the operation of its own wires, the Western Union was supplying customers with various kinds of printing-telegraphs and dial telegraphs, some of which could transmit sixty words a minute. These accurate instruments, it believed, could never be displaced by such a scientific oddity as the telephone. And it continued to believe this until one of its subsidiary companies - the Gold and Stock - reported that several of its machines had been superseded by telephones.

At once the Western Union awoke from its indifference. Even this tiny nibbling at its business must be stopped. It took action quickly and organized the "American Speaking-Telephone Company," with $300,000 capital, and with three electrical inventors, Edison, Gray, and Dolbear, on its staff. With all the bulk of its great wealth and prestige, it swept down upon Bell and his little bodyguard. It trampled upon Bell's patent with as little concern as an elephant can have when he tramples upon an ant's nest. To the complete bewilderment of Bell, it coolly announced that it had "the only original telephone," and that it was ready to supply "superior telephones with all the latest improvements made by the original inventors - Dolbear, Gray, and Edison."

The result was strange and unexpected. The Bell group, instead of being driven from the field, were at once lifted to a higher level in the business world. The effect was as if the Standard Oil Company were to commence the manufacture of aeroplanes. In a flash, the telephone ceased to be a "scientific toy," and became an article of commerce. It began for the first time to be taken seriously. And the Western Union, in the endeavor to protect its private lines, became involuntarily a bell-wether to lead capitalists in the direction of the telephone.

Sanders's relatives, who were many and rich, came to his rescue. Most of them were well-known business men - the Bradleys, the Saltonstalls, Fay, Silsbee, and Carlton. These men, together with Colonel William H. Forbes, who came in as a friend of the Bradleys, were the first capitalists who, for purely business reasons, invested money in the Bell patents. Two months after the Western Union had given its weighty endorsement to the telephone, these men organized a company to do business in New England only, and put fifty thousand dollars in its treasury.

In a short time the delighted Hubbard found himself leasing telephones at the rate of a thousand a month. He was no longer a promoter, but a general manager. Men were standing in line to ask for agencies. Crude little telephone exchanges were being started in a dozen or more cities. There was a spirit of confidence and enterprise; and the next step, clearly, was to create a business organization. None of the partners were competent to undertake such a work. Hubbard had little aptitude as an organizer; Bell had none; and Sanders was held fast by his leather interests. Here, at last, after four years of the most heroic effort, were the raw materials out of which a telephone business could be constructed. But who was to be the builder, and where was he to be found?

One morning the indefatigable Hubbard solved the problem. "Watson," he said, "there's a young man in Washington who can handle this situation, and I want you to run down and see what you think of him." Watson went, reported favorably, and in a day or so the young man received a letter from Hubbard, offering him the position of General Manager, at a salary of thirty-five hundred dollars a year. "We rely,"

Hubbard said, "upon your executive ability, your fidelity, and unremitting zeal." The young man replied, in one of those dignified letters more usual in the nineteenth than in the twentieth century. "My faith in the success of the enterprise is such that I am willing to trust to it," he wrote, "and I have confidence that we shall establish the harmony and cooperation that is essential to the success of an enterprise of this kind." One week later the young man, Theodore N. Vail, took his seat as General Manager in a tiny office in Reade Street, New York, and the building of the business began.

This arrival of Vail at the critical moment emphasized the fact that Bell was one of the most fortunate of inventors. He was not robbed of his invention, as might easily have happened. One by one there arrived to help him a number of able men, with all the various abilities that the changing situation required. There was such a focussing of factors that the whole matter appeared to have been previously rehearsed. No sooner had Bell appeared on the stage than his supporting players, each in his turn, received his cue and took part in the action of the drama. There was not one of these men who could have done the work of any other. Each was distinctive and indispensable. Bell invented the telephone; Watson constructed it; Sanders financed it; Hubbard introduced it; and Vail put it on a business basis.

The new General Manager had, of course, no experience in the telephone business. Neither had any one else. But he, like Bell, came to his task with a most surprising fitness. He was a member of the historic Vail family of Morristown, New Jersey, which had operated the Speedwell Iron Works for four or five generations. His grand-uncle Stephen had built the engines for the Savannah, the first American steamship to cross the Atlantic Ocean; and his cousin Alfred was the friend and co-worker of Morse, the inventor of the telegraph. Morse had lived for several years at the Vail homestead in Morristown; and it was here that he erected his first telegraph line, a three-mile circle around the Iron Works, in 1838. He and Alfred Vail experimented side by side in the making of the telegraph, and Vail eventually received a fortune for his share of the Morse patent.

Thus it happened that young Theodore Vail learned the dramatic story of Morse at his mother's knee. As a boy, he played around the first telegraph line, and learned to put messages on the wire. His favorite toy was a little telegraph that he constructed for himself. At twenty-two he went West, in the vague hope of possessing a bonanza farm; then he swung back into telegraphy, and in a few years found himself in the Government Mail Service at Washington. By 1876, he was at the head of this Department, which he completely reorganized. He introduced the bag system in postal cars, and made war on waste and clumsiness. By virtue of this position he was the one man in the United States who had a comprehensive view of all railways and telegraphs. He was much more apt, consequently, than other men to develop the idea of a national telephone system.

While in the midst of this bureaucratic house-cleaning he met Hubbard, who had just been appointed by President Hayes as the head of a commission on mail transportation. He and Hubbard were constantly thrown together, on trains and in hotels; and as Hubbard invariably had a pair of telephones in his valise, the two men soon became co-enthusiasts. Vail found himself painting brain-pictures of the future of the telephone, and by the time that he was asked to become its General Manager, he had become so confident that, as he said afterwards, he "was willing to leave a Government job with a small salary for a telephone job with no salary."

So, just as Amos Kendall had left the post office service thirty years before to establish the telegraph business, Theodore N. Vail left the post office service to establish the telephone business. He had been in authority over thirty-five hundred postal employees, and was the developer of a system that covered every inhabited portion of the country. Consequently, he had a quality of experience that was immensely valuable in straightening out the tangled affairs of the telephone. Line by line, he mapped out a method, a policy, a system. He introduced a larger view of the telephone business, and swept off the table all schemes for selling out. He persuaded half a dozen of his post office friends to buy stock, so that in less than two months the first "Bell Telephone Company" was organized, with $450,000 capital and a service of twelve thousand telephones.

Vail's first step, naturally, was to stiffen up the backbone of this little company, and to prevent the Western Union from frightening it into a surrender. He immediately sent a copy of Bell's patent to every agent, with orders to hold the fort against all opposition. "We have the only original telephone patents," he wrote; "we have organized and introduced the business, and we do not propose to have it taken from us by any corporation." To one agent, who was showing the white feather, he wrote:

"You have too great an idea of the Western Union. If it was all massed in your one city you might well fear it; but it is represented there by one man only, and he has probably as much as he can attend to outside of the telephone. For you to acknowledge that you cannot compete with his influence when you make it your special business, is hardly the thing. There may be a dozen concerns that will all go to the Western Union, but they will not take with them all their friends. I would advise that you go ahead and keep your present advantage. We must organize companies with sufficient vitality to carry on a fight, as it is simply useless to get a company started that will succumb to the first bit of opposition it may encounter."

Next, having encouraged his thoroughly alarmed agents, Vail proceeded to build up a definite business policy. He stiffened up the contracts and made them good for five years only. He confined each agent to one place, and reserved all rights to connect one city with another. He established a department to collect and protect any new inventions that concerned the telephone. He agreed to take part of the royalties in stock, when any local company preferred to pay its debts in this way. And he took steps toward standardizing all telephonic apparatus by controlling the factories that made it.

These various measures were part of Vail's plan to create a national telephone system. His central idea, from the first, was not the mere leasing of telephones, but rather the creation of a Federal company that would be a permanent partner in the entire telephone business. Even in that day of small things, and amidst the confusion and rough-and-tumble of pioneering, he worked out the broad policy that prevails to-day; and this goes far to explain the fact that there are in the United States twice as many telephones as there are in all other countries combined.

Vail arrived very much as Blucher did at the battle of Waterloo - a trifle late, but in time to prevent the telephone forces from being routed by the Old Guard of the Western Union. He was scarcely seated in his managerial chair, when the Western Union threw the entire Bell army into confusion by launching the Edison transmitter. Edison, who was at that time fairly started in his career of wizardry, had made an instrument of marvellous alertness. It was beyond all argument superior to the telephones then in use and the lessees of Bell telephones clamored with one voice for "a transmitter as good as Edison's." This, of course, could not be had in a moment, and the five months that followed were the darkest days in the childhood of the telephone.

How to compete with the Western Union, which had this superior transmitter, a host of agents, a network of wires, forty millions of capital, and a first claim upon all newspapers, hotels, railroads, and rights of way - that was the immediate problem that confronted the new General Manager. Every inch of progress had to be fought for. Several of his captains deserted, and he was compelled to take control of their unprofitable exchanges. There was scarcely a mail that did not bring him some bulletin of discouragement or defeat.

In the effort to conciliate a hostile public, the telephone rates had everywhere been made too low. Hubbard had set a price of twenty dollars a year, for the use of two telephones on a private line; and when exchanges were started, the rate was seldom more than three dollars a month. There were deadheads in abundance, mostly officials and politicians. In St. Louis, one of the few cities that charged a sufficient price, nine-tenths of the merchants refused to become subscribers. In Boston, the first pay-station ran three months before it earned a dollar. Even as late as 1880, when the first National Telephone Convention was held at Niagara Falls, one of the delegates expressed the general situation very correctly when he said: "We were all in a state of enthusiastic uncertainty. We were full of hope, yet when we analyzed those hopes they were very airy indeed. There was probably not one company that could say it was making a cent, nor even that it EXPECTED to make a cent."

Especially in the largest cities, where the Western Union had most power, the lives of the telephone pioneers were packed with hardships and adventures. In Philadelphia, for instance, a resolute young man named Thomas E. Cornish was attacked as though he had suddenly become a public enemy, when he set out to establish the first telephone service. No official would grant him a permit to string wires. His workmen were arrested. The printing-telegraph men warned him that he must either quit or be driven out. When he asked capitalists for money, they replied that he might as well expect to lease jew's-harps as telephones. Finally, he was compelled to resort to strategy where argument had failed. He had received an order from Colonel Thomas Scott, who wanted a wire between his house and his office. Colonel Scott was the President of the Pennsylvania Railroad, and therefore a man of the highest prestige in the city. So as soon as Cornish had put this line in place, he kept his men at work stringing other lines. When the police interfered, he showed them Colonel Scott's signature and was let alone. In this way he put fifteen wires up before the trick was discovered; and soon afterwards, with eight subscribers, he founded the first Philadelphia exchange.

As may be imagined, such battling as this did not put much money into the treasury of the parent company; and the letters written by Sanders at this time prove that it was in a hard plight.

The following was one of the queries put to Hubbard by the overburdened Sanders:

"How on earth do you expect me to meet a draft of two hundred and seventy-five dollars without a dollar in the treasury, and with a debt of thirty thousand dollars staring us in the face?" "Vail's salary is small enough," he continued in a second letter, "but as to where it is coming from I am not so clear. Bradley is awfully blue and discouraged. Williams is tormenting me for money and my personal credit will not stand everything. I have advanced the Company two thousand dollars to-day, and Williams must have three thousand dollars more this month. His pay-day has come and his capital will not carry him another inch. If Bradley throws up his hand, I will unfold to you my last desperate plan."

And if the company had little money, it had less credit. Once when Vail had ordered a small bill of goods from a merchant named Tillotson, of 15 Dey Street, New York, the merchant replied that the goods were ready, and so was the bill, which was seven dollars. By a strange coincidence, the magnificent building of the New York Telephone Company stands to-day on the site of Tillotson's store.

Month after month, the little Bell Company lived from hand to mouth. No salaries were paid in full. Often, for weeks, they were not paid at all. In Watson's note-book there are such entries during this period as "Lent Bell fifty cents," "Lent Hubbard twenty cents," "Bought one bottle beer - too bad can't have beer every day." More than once Hubbard would have gone hungry had not Devonshire, the only clerk, shared with him the contents of a dinner-pail. Each one of the little group was beset by taunts and temptations. Watson was offered ten thousand dollars for his one-tenth interest, and hesitated three days before refusing it. Railroad companies offered Vail a salary that was higher and sure, if he would superintend their mail business. And as for Sanders, his folly was the talk of Haverhill. One Haverhill capitalist, E. J. M. Hale, stopped him on the street and asked, "Have n't you got a good leather business, Mr. Sanders?" "Yes," replied Sanders. "Well," said Hale, "you had better attend to it and quit playing on wind instruments." Sanders's banker, too, became uneasy on one occasion and requested him to call at the bank. "Mr. Sanders," he said, "I will be obliged if you will take that telephone stock out of the bank, and give me in its place your note for thirty thousand dollars. I am expecting the examiner here in a few days, and I don't want to get caught with that stuff in the bank."

Then, in the very midnight of this depression, poor Bell returned from England, whither he and his bride had gone on their honeymoon, and announced that he had no money; that he had failed to establish a telephone business in England; and that he must have a thousand dollars at once to pay his urgent debts. He was thoroughly discouraged and sick. As he lay in the Massachusetts General Hospital, he wrote a cry for help to the embattled little company that was making its desperate fight to protect his patents. "Thousands of telephones are now in operation in all parts of the country," he said, "yet I have not yet received one cent from my invention. On the contrary, I am largely out of pocket by my researches, as the mere value of the profession that I have sacrificed during my three years' work, amounts to twelve thousand dollars."

Fortunately, there came, in almost the same mail with Bell's letter, another letter from a young Bostonian named Francis Blake, with the good news that he had invented a transmitter as satisfactory as Edison's, and that he would prefer to sell it for stock instead of cash. If ever a man came as an angel of light, that man was Francis Blake. The possession of his transmitter instantly put the Bell Company on an even footing with the Western Union, in the matter of apparatus. It encouraged the few capitalists who had invested money, and it stirred others to come forward. The general business situation had by this time become more settled, and in four months the company had twenty-two thousand telephones in use, and had reorganized into the National Bell Telephone Company, with $850, 000 capital and with Colonel Forbes as its first President. Forbes now picked up the load that had been carried so long by Sanders. As the son of an East India merchant and the son-in-law of Ralph Waldo Emerson, he was a Bostonian of the Brahmin caste. He was a big, four-square man who was both popular and efficient; and his leadership at this crisis was of immense value.

This reorganization put the telephone business into the hands of competent business men at every point. It brought the heroic and experimental period to an end.

From this time onwards the telephone had strong friends in the financial world. It was being attacked by the Western Union and by rival inventors who were jealous of Bell's achievement. It was being half-starved by cheap rates and crippled by clumsy apparatus. It was being abused and grumbled at by an impatient public. But the art of making and marketing it had at last been built up into a commercial enterprise. It was now a business, fighting for its life.

CHAPTER III

THE HOLDING OF THE BUSINESS

For seventeen months no one disputed Bell's claim to be the original inventor of the telephone. All the honor, such as it was, had been given to him freely, and no one came forward to say that it was not rightfully his. No one, so far as we know, had any strong desire to do so. No one conceived that the telephone would ever be any more than a whimsical oddity of science. It was so new, so unexpected, that from Lord Kelvin down to the messenger boys in the telegraph offices, it was an incomprehensible surprise. But after Bell had explained his invention in public lectures before more than twenty thousand people, after it had been on exhibition for months at the Philadelphia Centennial, after several hundred articles on it had appeared in newspapers and scientific magazines, and after actual sales of telephones had been made in various parts of the country, there began to appear such a succession of claimants and infringers that the forgetful public came to believe that the telephone, like most inventions, was the product of many minds.

Just as Morse, who was the sole inventor of the American telegraph in 1837, was confronted by sixty-two rivals in 1838, so Bell, who was the sole inventor in 1876, found himself two years later almost mobbed by the "Tichborne claimants" of the telephone. The inventors who had been his competitors in the attempt to produce a musical telegraph, persuaded themselves that they had unconsciously done as much as he. Any possessor of a telegraphic patent, who had used the common phrase "talking wire," had a chance to build up a plausible story of prior invention. And others came forward with claims so vague and elusive that Bell would scarcely have been more surprised if the heirs of Goethe had demanded a share of the telephone royalties on the ground that Faust had spoken of "making a bridge through the moving air."

This babel of inventors and pretenders amazed Bell and disconcerted his backers. But it was no more than might have been expected. Here was a patent - "the most valuable single patent ever issued" - and yet the invention itself was so simple that it could be duplicated easily by any smart boy or any ordinary mechanic. The making of a telephone was like the trick of Columbus standing an egg on end. Nothing was easier to those who knew how. And so it happened that, as the crude little model of Bell's original telephone lay in the Patent Office open and unprotected except by a few phrases that clever lawyers might evade, there sprang up inevitably around it the most costly and persistent Patent War that any country has ever known, continuing for eleven years and comprising SIX HUNDRED LAWSUITS.

The first attack upon the young telephone business was made by the Western Union Telegraph Company. It came charging full tilt upon Bell, driving three inventors abreast - Edison, Gray, and Dolbear. It expected an easy victory; in fact, the disparity between the two opponents was so evident, that there seemed little chance of a contest of any kind. "The Western Union will swallow up the telephone people," said public opinion, "just as it has already swallowed up all improvements in telegraphy."

At that time, it should be remembered, the Western Union was the only corporation that was national in its extent. It was the most powerful electrical company in the world, and, as Bell wrote to his parents, "probably the largest corporation that ever existed." It had behind it not only forty millions of capital, but the prestige of the Vanderbilts, and the favor of financiers everywhere. Also, it met the telephone pioneers at every point because it, too, was a WIRE company. It owned rights-of-way along roads and on house-tops. It had a monopoly of hotels and railroad offices. No matter in what direction the Bell Company turned, the live wire of the Western Union lay across its path.

From the first, the Western Union relied more upon its strength than upon the merits of its case. Its chief electrical expert, Frank L. Pope, had made a six months' examination of the Bell patents. He had bought every book in the United States and Europe that was likely to have any reference to the transmission of speech, and employed a professor who knew eight languages to translate them. He and his men ransacked libraries and patent offices; they rummaged and sleuthed and interviewed; and found nothing of any value. In his final report to the Western Union, Mr. Pope announced that there was no way to make a telephone except Bell's way, and advised the purchase of the Bell patents. "I am entirely unable to discover any apparatus or method anticipating the invention of Bell as a whole," he said; "and I conclude that his patent is valid." But the officials of the great corporation refused to take this report seriously. They threw it aside and employed Edison, Gray, and Dolbear to devise a telephone that could be put into competition with Bell's.

As we have seen in the previous chapter, there now came a period of violent competition which is remembered as the Dark Ages of the telephone business. The Western Union bought out several of the Bell exchanges and opened up a lively war on the others. As befitting its size, it claimed everything. It introduced Gray as the original inventor of the telephone, and ordered its lawyers to take action at once against the Bell Company for infringement of the Gray patent. This high-handed action, it hoped, would most quickly bring the little Bell group into a humble and submissive frame of mind. Every morning the Western Union looked to see the white flag flying over the Bell headquarters. But no white flag appeared. On the contrary, the news came that the Bell Company had secured two eminent lawyers and were ready to give battle.

The case began in the Autumn of 1878 and lasted for a year. Then it came to a sudden and most unexpected ending. The lawyer-in-chief of the Western Union was George Gifford, who was perhaps the ablest patent attorney of his day. He was versed in patent lore from Alpha to Omega; and as the trial proceeded, he became convinced that the Bell patent was valid. He notified the Western Union confidentially, of course, that its case could not be proven, and that "Bell was the original inventor of the telephone." The best policy, he suggested, was to withdraw their claims and make a settlement. This wise advice was accepted, and the next day the white flag was hauled up, not by the little group of Bell fighters, who were huddled together in a tiny, two-room office, but by the mighty Western Union itself, which had been so arrogant when the encounter began.

A committee of three from each side was appointed, and after months of disputation, a treaty of peace was drawn up and signed. By the terms of this treaty the Western Union agreed -

(1) To admit that Bell was the original inventor.

(2) To admit that his patents were valid.

(3) To retire from the telephone business.

The Bell Company, in return for this surrender, agreed

(1) To buy the Western Union telephone system.

(2) To pay the Western Union a royalty of twenty per cent on all telephone rentals.

(3) To keep out of the telegraph business.

This agreement, which was to remain in force for seventeen years, was a master-stroke of diplomacy on the part of the Bell Company. It was the Magna Charta of the telephone. It transformed a giant competitor into a friend. It added to the Bell System fifty-six thousand telephones in fifty-five cities. And it swung the valiant little company up to such a pinnacle of prosperity that its stock went skyrocketing until it touched one thousand dollars a share.

The Western Union had lost its case, for several very simple reasons: It had tried to operate a telephone system on telegraphic lines, a plan that has invariably been unsuccessful, it had a low idea of the possibilities of the telephone business; and its already busy agents had little time or knowledge or enthusiasm to give to the new enterprise. With all its power, it found itself outfought by this compact body of picked men, who were young, zealous, well-handled, and protected by a most invulnerable patent.

The Bell Telephone now took its place with the Telegraph, the Railroad, the Steamboat, the Harvester, and the other necessities of a civilized country. Its pioneer days were over. There was no more ridicule and incredulity. Every one knew that the Bell people had whipped the Western Union, and hastened to join in the grand Te Deum of applause. Within five months from the signing of the agreement, there had to be a reorganization; and the American Bell Telephone Company was created, with six million dollars capital. In the following year, 1881, twelve hundred new towns and cities were marked on the telephone map, and the first dividends were paid - $178,500. And in 1882 there came such a telephone boom that the Bell System was multiplied by two, with more than a million dollars of gross earnings.

At this point all the earliest pioneers of the telephone, except Vail, pass out of its history. Thomas Sanders sold his stock for somewhat less than a million dollars, and presently lost most of it in a Colorado gold mine. His mother, who had been so good a friend to Bell, had her fortune doubled. Gardiner G. Hubbard withdrew from business life, and as it was impossible for a man of his ardent temperament to be idle, he plunged into the National Geographical Society. He was a Colonel Sellers whose dream of millions (for the telephone) had come true; and when he died, in 1897, he was rich both in money and in the affection of his friends. Charles Williams, in whose workshop the first telephones were made, sold his factory to the Bell Company in 1881 for more money than he had ever expected to possess. Thomas A. Watson resigned at the same time, finding himself no longer a wage-worker but a millionaire. Several years later he established a shipbuilding plant near Boston, which grew until it employed four thousand workmen and had built half a dozen warships for the United States Navy.

As for Bell, the first cause of the telephone business, he did what a true scientific Bohemian might have been expected to do; he gave all his stock to his bride on their marriage-day and resumed his work as an instructor of deaf-mutes. Few kings, if any, had ever given so rich a wedding present; and certainly no one in any country ever obtained and tossed aside an immense fortune as incidentally as did Bell. When the Bell Company offered him a salary of ten thousand dollars a year to remain its chief inventor, he refused the offer cheerfully on the ground that he could not "invent to order." In 1880, the French Government gave him the Volta Prize of fifty thousand francs and the Cross of the Legion of Honor. He has had many honors since then, and many interests. He has been for thirty years one of the most brilliant and picturesque personalities in American public life. But none of his later achievements can in any degree compare with what he did in a cellar in Salem, at twenty-eight years of age.

They had all become rich, these first friends of the

telephone, but not fabulously so. There was not at that time, nor has there been since, any one who became a multimillionaire by the sale of telephone service. If the Bell Company had sold its stock at the highest price reached, in 1880, it would have received less than nine million dollars - a huge sum, but not too much to pay for the invention of the telephone and the building up of a new art and a new industry. It was not as much as the value of the eggs laid during the last twelve months by the hens of Iowa.

But, as may be imagined, when the news of the Western Union agreement became known, the story of the telephone became a fairy tale of success. Theodore Vail was given a banquet by his old-time friends in the Washington postal service, and toasted as "the Monte Cristo of the Telephone." It was said that the actual cost of the Bell plant was only one-twenty-fifth of its capital, and that every four cents of investment had thus become a dollar. Even Jay Gould, carried beyond his usual caution by these stories, ran up to New Haven and bought its telephone company, only to find out later that its earnings were less than its expenses.

Much to the bewilderment of the Bell Company, it soon learned that the troubles of wealth are as numerous as those of poverty. It was beset by a throng of promoters and stock-jobbers, who fell upon it and upon the public like a swarm of seventeen-year locusts. In three years, one hundred and twenty-five competing companies were started, in open defiance of the Bell patents. The main object of these companies was not, like that of the Western Union, to do a legitimate telephone business, but to sell stock to the public. The face value of their stock was $225,000,000, although few of them ever sent a message. One company of unusual impertinence, without money or patents, had capitalized its audacity at $15,000,000.

How to HOLD the business that had been established - that was now the problem. None of the Bell partners had been mere stock-jobbers. At one time they had even taken a pledge not to sell any of their stock to outsiders. They had financed their company in a most honest and simple way; and they were desperately opposed to the financial banditti whose purpose was to transform the telephone business into a cheat and a gamble. At first, having held their own against the Western Union, they expected to make short work of the stock-jobbers. But it was a vain hope. These bogus companies, they found, did not fight in the open, as the Western Union had done.

All manner of injurious rumors were presently set afloat concerning the Bell patent. Other inventors - some of them honest men, and some shameless pretenders - were brought forward with strangely concocted tales of prior invention. The Granger movement was at that time a strong political factor in the Middle West, and its blind fear of patents and "monopolies" was turned aggressively against the Bell Company. A few Senators and legitimate capitalists were lifted up as the figureheads of the crusade. And a loud hue-and-cry was raised in the newspapers against "high rates and monopoly" to distract the minds of the people from the real issue of legitimate business versus stock-company bubbles.

The most plausible and persistent of all the various inventors who snatched at Bell's laurels, was Elisha Gray. He refused to abide by the adverse decision of the court. Several years after his defeat, he came forward with new weapons and new methods of attack. He became more hostile and irreconcilable; and until his death, in 1901, never renounced his claim to be the original inventor of the telephone.

The reason for this persistence is very evident. Gray was a professional inventor, a highly competent man who had begun his career as a blacksmith's apprentice, and risen to be a professor of Oberlin. He made, during his lifetime, over five million dollars by his patents. In 1874, he and Bell were running a neck-and-neck race to see who could first invent a musical telegraph - when, presto! Bell suddenly turned aside, because of his acoustical knowledge, and invented the telephone, while Gray kept straight ahead. Like all others who were in quest of a better telegraph instrument, Gray had glimmerings of the possibility of sending speech by wire, and by one of the strangest of coincidences he filed a caveat on the subject on the SAME DAY that Bell filed the application for a patent. Bell had arrived first. As the record book shows, the fifth entry on that day was: "A. G. Bell, $15"; and the thirty-ninth entry was "E. Gray, $10."

There was a vast difference between Gray's caveat and Bell's application. A caveat is a declaration that the writer has NOT invented a thing, but believes that he is about to do so; while an APPLICATION is a declaration that the writer has already perfected the invention. But Gray could never forget that he had seemed to be, for a time, so close to the golden prize; and seven years after he had been set aside by the Western Union agreement, he reappeared with claims that had grown larger and more definite.

When all the evidence in the various Gray lawsuits is

sifted out, there appear to have been three distinctly different Grays: first, Gray the SCOFFER, who examined Bell's telephone at the Centennial and said it was "nothing but the old lover's telegraph. It is impossible to make a practical speaking telephone on the principle shown by Professor Bell.... The currents are too feeble"; second, Gray the CONVERT, who wrote frankly to Bell in 1877, "I do not claim the credit of inventing it"; and third, Gray the CLAIMANT, who endeavored to prove in 1886 that he was the original inventor. His real position in the matter was once well and wittily described by his partner, Enos M. Barton, who said: "Of all the men who DIDN'T invent the telephone, Gray was the nearest."

It is now clearly seen that the telephone owes nothing to Gray. There are no Gray telephones in use in any country. Even Gray himself, as he admitted in court, failed when he tried to make a telephone on the lines laid down in his caveat. The final word on the whole matter was recently spoken by George C. Maynard, who established the telephone business in the city of Washington. Said Mr. Maynard:

"Mr. Gray was an intimate and valued friend of mine, but it is no disrespect to his memory to say that on some points involved in the telephone matter, he was mistaken. No subject was ever so thoroughly investigated as the invention of the speaking telephone. No patent has ever been submitted to such determined assault from every direction as Bell's; and no inventor has ever been more completely vindicated. Bell was the first inventor, and Gray was not."

After Gray, the weightiest challenger who came against Bell was Professor Amos E. Dolbear, of Tufts

College. He, like Gray, had written a letter of applause to Bell in 1877. "I congratulate you, sir," he said, "upon your very great invention, and I hope to see it supplant all forms of existing telegraphs, and that you will be successful in obtaining the wealth and honor which is your due." But one year later, Dolbear came to view with an opposition telephone. It was not an imitation of Bell's, he insisted, but an improvement upon an electrical device made by a German named Philip Reis, in 1861.

Thus there appeared upon the scene the so-called "Reis telephone," which was not a telephone at all, in any practical sense, but which served well enough for nine years or more as a weapon to use against the Bell patents. Poor Philip Reis himself, the son of a baker in Frankfort, Germany, had hoped to make a telephone, but he had failed. His machine was operated by a "make-and-break" current, and so could not carry the infinitely delicate vibrations made by the human voice. It could transmit the pitch of a sound, but not the QUALITY. At its best, it could carry a tune, but never at any time a spoken sentence. Reis, in his later years, realized that his machine could never be used for the transmission of conversation; and in a letter to a friend he tells of a code of signals that he has invented.

Bell had once, during his three years of experimenting, made a Reis machine, although at that time he had not seen one. But he soon threw it aside, as of no practical value. As a teacher of acoustics, Bell knew that the one indispensable requirement of a telephone is that it shall transmit the WHOLE of a sound, and not merely the pitch of it. Such scientists as Lord Kelvin, Joseph Henry, and Edison had seen the little Reis instrument years before Bell invented the telephone; but they regarded it as a mere musical toy. It was "not in any sense a speaking telephone," said Lord Kelvin. And Edison, when trying to put the Reis machine in the most favorable light, admitted humorously that when he used a Reis transmitter he generally "knew what was coming; and knowing what was coming, even a Reis transmitter, pure and simple, reproduces sounds which seem almost like that which was being transmitted; but when the man at the other end did not know what was coming, it was very seldom that any word was recognized."

In the course of the Dolbear lawsuit, a Reis machine was brought into court, and created much amusement. It was able to squeak, but not to speak. Experts and professors wrestled with it in vain. It refused to transmit one intelligible sentence. "It CAN speak, but it WON'T," explained one of Dolbear's lawyers. It is now generally known that while a Reis machine, when clogged and out of order, would transmit a word or two in an imperfect way, it was built on wrong lines. It was no more a telephone than a wagon is a sleigh, even though it is possible to chain the wheels and make them slide for a foot or two. Said Judge Lowell, in rendering his famous decision:

"A century of Reis would never have produced a speaking telephone by mere improvement of construction. It was left for Bell to discover that the failure was due not to workmanship but to the principle which was adopted as the basis of what had to be done. . . . Bell discovered a new art - that of transmitting speech by electricity, and his claim is not as broad as his invention. . . . To follow Reis is to fail; but to follow Bell is to succeed."

After the victory over Dolbear, the Bell stock went soaring skywards; and the higher it went, the greater were the number of infringers and blowers of stock bubbles. To bait the Bell Company became almost a national sport. Any sort of claimant, with any sort of wild tale of prior invention, could find a speculator to support him. On they came, a motley array, "some in rags, some on nags, and some in velvet gowns." One of them claimed to have done wonders with an iron hoop and a file in 1867; a second had a marvellous table with glass legs; a third swore that he had made a telephone in 1860, but did not know what it was until he saw Bell's patent; and a fourth told a vivid story of having heard a bullfrog croak via a telegraph wire which was strung into a certain cellar in Racine, in 1851.

This comic opera phase came to a head in the famous Drawbaugh case, which lasted for nearly four years, and filled ten thousand pages with its evidence. Having failed on Reis, the German, the opponents of Bell now brought forward an American inventor named Daniel Drawbaugh, and opened up a noisy newspaper campaign. To secure public sympathy for Drawbaugh, it was said that he had invented a complete telephone and switchboard before 1876, but was in such "utter and abject poverty" that he could not get himself a patent. Five hundred witnesses were examined; and such a general turmoil was aroused that the Bell lawyers were compelled to take the attack seriously, and to fight back with every pound of ammunition they possessed.

The fact about Drawbaugh is that he was a mechanic in a country village near Harrisburg, Pennsylvania. He was ingenious but not inventive; and loved to display his mechanical skill before the farmers and villagers. He was a subscriber to The Scientific American; and it had become the fixed habit of his life to copy other people's inventions and exhibit them as his own. He was a trailer of inventors. More than forty instances of this imitative habit were shown at the trial, and he was severely scored by the judge, who accused him of "deliberately falsifying the facts." His ruling passion of imitation, apparently, was not diminished by the loss of his telephone claims, as he came to public view again in 1903 as a trailer of Marconi.

Drawbaugh's defeat sent the Bell stock up once more, and brought on a Xerxes' army of opposition which called itself the "Overland Company." Having learned that no one claimant could beat Bell in the courts, this company massed the losers together and came forward with a scrap-basket full of patents. Several powerful capitalists undertook to pay the expenses of this adventure. Wires were strung; stock was sold; and the enterprise looked for a time so genuine that when the Bell lawyers asked for an injunction against it, they were refused. This was as hard a blow as the Bell people received in their eleven years of litigation; and the Bell stock tumbled thirty-five points in a few days. Infringing companies sprang up like gourds in the night. And all went merrily with the promoters until the Overland Company was thrown out of court, as having no evidence, except "the refuse and dregs of former cases - the heel-taps found in the glasses at the end of the frolic."

But even after this defeat for the claimants, the frolic was not wholly ended. They next planned to get through politics what they could not get through law; they induced the Government to bring suit for the annulment of the Bell patents. It was a bold and desperate move, and enabled the promoters of paper companies to sell stock for several years longer. The whole dispute was re-opened, from Gray to Drawbaugh. Every battle was re-fought; and in the end, of course, the Government officials learned that they were being used to pull telephone chestnuts out of the fire. The case was allowed to die a natural death, and was informally dropped in 1896.

In all, the Bell Company fought out thirteen lawsuits that were of national interest, and five that were carried to the Supreme Court in Washington. It fought out five hundred and eighty-seven other lawsuits of various natures; and with the exception of two trivial contract suits, IT NEVER LOST A CASE.

Its experience is an unanswerable indictment of our system of protecting inventors. No inventor had ever a clearer title than Bell. The Patent Office itself, in 1884, made an eighteen-months' investigation of all telephone patents, and reported: "It is to Bell that the world owes the possession of the speaking telephone." Yet his patent was continuously under fire, and never at any time secure. Stock companies whose paper capital totalled more than $500,000,000 were organized to break it down; and from first to last the success of the telephone was based much less upon the monopoly of patents than upon the building up of a well organized business.

Fortunately for Bell and the men who upheld him, they were defended by two master-lawyers who have seldom, if ever, had an equal for team work and efficiency - Chauncy Smith and James J. Storrow. These two men were marvellously well mated. Smith was an old-fashioned attorney of the Websterian sort, dignified, ponderous, and impressive. By 1878, when he came in to defend the little Bell Company against the towering Western Union, Smith had become the most noted patent lawyer in Boston. He was a large, thick-set man, a reminder of Benjamin Franklin, with clean-shaven face, long hair curling at the ends, frock coat, high collar, and beaver hat.

Storrow, on the contrary, was a small man, quiet in manner, conversational in argument, and an encyclopedia of definite information. He was so thorough that, when he became a Bell lawyer, he first spent an entire summer at his country home in Petersham, studying the laws of physics and electricity. He was never in the slightest degree spectacular. Once only, during the eleven years of litigation, did he lose control of his temper. He was attacking the credibility of a witness whom he had put on the stand, but who had been tampered with by the opposition lawyers. "But this man is your own witness," protested the lawyers. "Yes," shouted the usually soft-speaking Storrow; "he WAS my witness, but now he is YOUR LIAR."

The efficiency of these two men was greatly increased by a third - Thomas D. Lockwood, who was chosen by Vail in 1879 to establish a Patent Department. Two years before, Lockwood had heard Bell lecture in Chickering Hall, New York, and was a "doubting Thomas." But a closer study of the telephone transformed him into an enthusiast. Having a memory like a filing system, and a knack for invention, Lockwood was well fitted to create such a department. He was a man born for the place. And he has seen the number of electrical patents grow from a few hundred in 1878 to eighty thousand in 1910.

These three men were the defenders of the Bell patents. As Vail built up the young telephone business, they held it from being torn to shreds in an orgy of speculative competition. Smith prepared the comprehensive plan of defence. By his sagacity and experience he was enabled to mark out the general principles upon which Bell had a right to stand. Usually, he closed the case, and he was immensely effective as he would declaim, in his deep voice: "I submit, Your Honor, that the literature of the world does not afford a passage which states how the human voice can be electrically transmitted, previous to the patent of Mr. Bell." His death, like his life, was dramatic. He was on his feet in the courtroom, battling against an infringer, when, in the middle of a sentence, he fell to the floor, overcome by sickness and the responsibilities he had carried for twelve years. Storrow, in a different way, was fully as indispensable as Smith. It was he who built up the superstructure of the Bell defence. He was a master of details. His brain was keen and incisive; and some of his briefs will be studied as long as the art of telephony exists. He might fairly have been compared, in action, to a rapid-firing Gatling gun; while Smith was a hundred-ton cannon, and Lockwood was the maker of the ammunition.

Smith and Storrow had three main arguments that never were, and never could be, answered. Fifty or more of the most eminent lawyers of that day tried to demolish these arguments, and failed. The first was Bell's clear, straightforward story of HOW HE DID IT, which rebuked and confounded the mob of pretenders. The second was the historical fact that the most eminent electrical scientists of Europe and America had seen Bell's telephone at the Centennial and had declared it to be NEW - "not only new but marvellous," said Tyndall. And the third was the very significant fact that no one challenged Bell's claim to be the original inventor of the telephone until his patent was seventeen months old.

The patent itself, too, was a remarkable document. It was a Gibraltar of security to the Bell Company. For eleven years it was attacked from all sides, and never dented. It covered an entire art, yet it was sustained during its whole lifetime. Printed in full, it would make ten pages of this book; but the core of it is in the last sentence: "The method of, and apparatus for, transmitting vocal or other sounds telegraphically, by causing electrical undulations, similar in form to the vibrations of the air accompanying the said vocal or other sounds." These words expressed an idea that had never been written before. It could not be evaded or overcome. There were only thirty-two words, but in six years these words represented an investment of a million dollars apiece.

Now that the clamor of this great patent war has died away, it is evident that Bell received no more credit and no more reward than he deserved. There was no telephone until he made one, and since he made one, no one has found out any other way. Hundreds of clever men have been trying for more than thirty years to outrival Bell, and yet every telephone in the world is still made on the plan that Bell discovered.

No inventor who preceded Bell did more, in the invention of the telephone, than to help Bell indirectly, in the same way that Fra Mauro and Toscanelli helped in the discovery of America by making the map and chart that were used by Columbus. Bell was helped by his father, who taught him the laws of acoustics; by

Helmholtz, who taught him the influence of magnets upon sound vibrations; by Koenig and Leon Scott, who taught him the infinite variety of these vibrations; by Dr. Clarence J. Blake, who gave him a human ear for his experiments; and by Joseph Henry and Sir Charles Wheatstone, who encouraged him to persevere. In a still more indirect way, he was helped by Morse's invention of the telegraph; by Faraday's discovery of the phenomena of magnetic induction; by Sturgeon's first electro-magnet; and by Volta's electric battery. All that scientists had achieved, from Galileo and Newton to Franklin and Simon Newcomb, helped Bell in a general way, by creating a scientific atmosphere and habit of thought. But in the actual making of the telephone, there was no one with Bell nor before him. He invented it first, and alone.

CHAPTER IV

THE DEVELOPMENT OF THE ART

Four wire-using businesses were already in the field when the telephone was born: the fire-alarm, burglar-alarm, telegraph, and messenger-boy service; and at first, as might have been expected, the humble little telephone was huddled in with these businesses as a sort of poor relation. To the general public, it was a mere scientific toy; but there were a few men, not many, in these wire-stringing trades, who saw a glimmering chance of creating a telephone business. They put telephones on the wires that were then in use. As these became popular, they added others. Each of their customers wished to be able to talk to every one else. And so, having undertaken to give telephone service, they presently found themselves battling with the most intricate and baffling engineering problem of modern times - the construction around the telephone of such a mechanism as would bring it into universal service.

The first of these men was Thomas A. Watson, the young mechanic who had been hired as Bell's helper. He began a work that to-day requires an army of twenty-six thousand people. He was for a couple of years the total engineering and manufacturing department of the telephone business, and by 1880 had taken out sixty patents for his own suggestions. It was Watson who took the telephone as Bell had made it, really a toy, with its diaphragm so delicate that a warm breath would put it out of order, and toughened it into a more rugged machine. Bell had used a disc of fragile gold-beaters' skin with a patch of sheet-iron glued to the centre. He could not believe, for a time, that a disc of all-iron would vibrate under the slight influence of a spoken word. But he and Watson noticed that when the patch was bigger the talking was better, and presently they threw away the gold-beaters' skin and used the iron alone.

Also, it was Watson who spent months experimenting with all sorts and sizes of iron discs, so as to get the one that would best convey the sound. If the iron was too thick, he discovered, the voice was shrilled into a Punch-and-Judy squeal; and if it was too thin, the voice became a hollow and sepulchral groan, as if the speaker had his head in a barrel. Other months, too, were spent in finding out the proper size and shape for the air cavity in front of the disc. And so, after the telephone had been perfected, IN PRINCIPLE, a full year was required to lift it out of the class of scientific toys, and another year or two to present it properly to the business world.

Until 1878 all Bell telephone apparatus was made by Watson in Charles Williams's little shop in Court Street, Boston - a building long since transformed into a five-cent theatre. But the business soon grew too big for the shop. Orders fell five weeks behind. Agents stormed and fretted. Some action had to be taken quickly, so licenses were given to four other manufacturers to make bells, switchboards, and so forth. By this time the Western Electric Company of

Chicago had begun to make the infringing Gray-Edison telephones for the Western Union, so that there were soon six groups of mechanics puzzling their wits over the new talk-machinery.

By 1880 there was plenty of telephonic apparatus being made, but in too many different varieties. Not all the summer gowns of that year presented more styles and fancies. The next step, if there was to be any degree of uniformity, was plainly to buy and consolidate these six companies; and by 1881 Vail had done this. It was the first merger in telephone history. It was a step of immense importance. Had it not been taken, the telephone business would have been torn into fragments by the civil wars between rival inventors.

From this time the Western Electric became the headquarters of telephonic apparatus. It was the Big Shop, all roads led to it. No matter where a new idea was born, sooner or later it came knocking at the door of the Western Electric to receive a material body. Here were the skilled workmen who became the hands of the telephone business. And here, too, were many of the ablest inventors and engineers, who did most to develop the cables and switchboards of to-day.

In Boston, Watson had resigned in 1882, and in his place, a year or two later stood a timely new arrival named E. T. Gilliland. This really notable man was a friend in need to the telephone. He had been a manufacturer of electrical apparatus in Indianapolis, until Vail's policy of consolidation drew him into the central group of pioneers and pathfinders. For five years Gilliland led the way as a developer of better and cheaper equipment. He made the best of a most difficult situation. He was so handy, so resourceful, that he invariably found a way to unravel the mechanical tangles that perplexed the first telephone agents, and this, too, without compelling them to spend large sums of capital. He took the ideas and apparatus that were then in existence, and used them to carry the telephone business through the most critical period of its life, when there was little time or money to risk on experiments. He took the peg switchboard of the telegraph, for instance, and developed it to its highest point, to a point that was not even imagined possible by any one else. It was the most practical and complete switchboard of its day, and held the field against all comers until it was superseded by the modern type of board, vastly more elaborate and expensive.

By 1884, gathered around Gilliland in Boston and the Western Electric in Chicago, there came to be a group of mechanics and high-school graduates, very young men, mostly, who had no reputations to lose; and who, partly for a living and mainly for a lark, plunged into the difficulties of this new business that had at that time little history and less prestige. These young adventurers, most of whom are still alive, became the makers of industrial history. They were unquestionably the founders of the present science of telephone engineering.

The problem that they dashed at so lightheartedly was much larger than any of them imagined. It was a Gibraltar of impossibilities. It was on the face of it a fantastic nightmare of a task - to weave such a web of wires, with interlocking centres, as would put any one telephone in touch with every other. There was no help for them in books or colleges. Watson, who had acquired a little knowledge, had become a shipbuilder.

Electrical engineering, as a profession, was unborn. And as for their telegraphic experience, while it certainly helped them for a time, it started them in the wrong direction and led them to do many things which had afterwards to be undone.

The peculiar electric current that these young pathfinders had to deal with is perhaps the quickest, feeblest, and most elusive force in the world. It is so amazing a thing that any description of it seems irrational. It is as gentle as a touch of a baby sunbeam, and as swift as the lightning flash. It is so small that the electric current of a single incandescent lamp is greater 500,000,000 times. Cool a spoonful of hot water just one degree, and the energy set free by the cooling will operate a telephone for ten thousand years. Catch the falling tear-drop of a child, and there will be sufficient water-power to carry a spoken message from one city to another.

Such is the tiny Genie of the Wire that had to be protected and trained into obedience. It was the most defenceless of all electric sprites, and it had so many enemies. Enemies! The world was populous with its enemies. There was the lightning, its elder brother, striking at it with murderous blows. There were the telegraphic and light-and-power currents, its strong and malicious cousins, chasing and assaulting it whenever it ventured too near. There were rain and sleet and snow and every sort of moisture, lying in wait to abduct it. There were rivers and trees and flecks of dust. It seemed as if all the known and unknown agencies of nature were in conspiracy to thwart or annihilate this gentle little messenger who had been conjured into life by the wizardry of Alexander Graham Bell.

All that these young men had received from Bell and Watson was that part of the telephone that we call the receiver. This was practically the sum total of Bell's invention, and remains to-day as he made it. It was then, and is yet, the most sensitive instrument that has ever been put to general use in any country. It opened up a new world of sound. It would echo the tramp of a fly that walked across a table, or repeat in New Orleans the prattle of a child in New York. This was what the young men received, and this was all. There were no switchboards of any account, no cables of any value, no wires that were in any sense adequate, no theory of tests or signals, no exchanges, NO TELEPHONE SYSTEM OF ANY SORT WHATEVER.

As for Bell's first telephone lines, they were as simple as clothes-lines. Each short little wire stood by itself, with one instrument at each end. There were no operators, switchboards, or exchanges. But there had now come a time when more than two persons wanted to be in the same conversational group. This was a larger use of the telephone; and while Bell himself had foreseen it, he had not worked out a plan whereby it could be carried out. Here was the new problem, and a most stupendous one - how to link together three telephones, or three hundred, or three thousand, or three million, so that any two of them could be joined at a moment's notice.

And that was not all. These young men had not only to battle against mystery and "the powers of the air"; they had not only to protect their tiny electric messenger, and to create a system of wire highways along which he could run up and down safely; they had to do more. They had to make this system so simple and fool-proof that every one - every one except the deaf and dumb -

could use it without any previous experience. They had to educate Bell's Genie of the Wire so that he would not only obey his masters, but anybody - anybody who could speak to him in any language.

No doubt, if the young men had stopped to consider their life-work as a whole, some of them might have turned back. But they had no time to philosophize. They were like the boy who learns how to swim by being pushed into deep water. Once the telephone business was started, it had to be kept going; and as it grew, there came one after another a series of congestions. Two courses were open; either the business had to be kept down to suit the apparatus, or the apparatus had to be developed to keep pace with the business. The telephone men, most of them, at least, chose development; and the brilliant inventions that afterwards made some of them famous were compelled by sheer necessity and desperation.

The first notable improvement upon Bell's invention was the making of the transmitter, in 1877, by Emile Berliner. This, too, was a romance. Berliner, as a poor German youth of nineteen, had landed in Castle Garden in 1870 to seek his fortune. He got a job as "a sort of bottle-washer at six dollars a week," he says, in a chemical shop in New York. At nights he studied science in the free classes of Cooper Union. Then a druggist named Engel gave him a copy of Muller's book on physics, which was precisely the stimulus needed by his creative brain. In 1876 he was fascinated by the telephone, and set out to construct one on a different plan. Several months later he had succeeded and was overjoyed to receive his first patent for a telephone transmitter. He had by this time climbed up from his bottle-washing to be a clerk in a drygoods store in Washington; but he was still poor and as unpractical as most inventors. Joseph Henry, the Sage of the American scientific world, was his friend, though too old to give him any help. Consequently, when Edison, two weeks later, also invented a transmitter, the prior claim of Berliner was for a time wholly ignored. Later the Bell Company bought Berliner's patent and took up his side of the case. There was a seemingly endless succession of delays - fourteen years of the most vexatious delays - until finally the Supreme Court of the United States ruled that Berliner, and not Edison, was the original inventor of the transmitter.

From first to last, the transmitter has been the product of several minds. Its basic idea is the varying of the electric current by varying the pressure between two points. Bell unquestionably suggested it in his famous patent, when he wrote of "increasing and diminishing the resistance." Berliner was the first actually to construct one. Edison greatly improved it by using soft carbon instead of a steel point. A Kentucky professor, David E. Hughes, started a new line of development by adapting a Bell telephone into a "microphone," a fantastic little instrument that would detect the noise made by a fly in walking across a table. Francis Blake, of Boston, changed a microphone into a practical transmitter. The Rev. Henry Hunnings, an English clergyman, hit upon the happy idea of using carbon in the form of small granules. And one of the Bell experts, named White, improved the Hunnings transmitter into its present shape. Both transmitter and receiver seem now to be as complete an artificial tongue and ear as human ingenuity can make them. They have persistently grown more elaborate, until today a telephone set, as it stands on a desk, contains as many as one hundred and thirty separate pieces, as well as a saltspoonful of glistening granules of carbon.

Next after the transmitter came the problem of the MYSTERIOUS NOISES. This was, perhaps, the most weird and mystifying of all the telephone problems. The fact was that the telephone had brought within hearing distance a new wonder-world of sound. All wires at that time were single, and ran into the earth at each end, making what was called a "grounded circuit." And this connection with the earth, which is really a big magnet, caused all manner of strange and uncouth noises on the telephone wires.

Noises! Such a jangle of meaningless noises had never been heard by human ears. There were spluttering and bubbling, jerking and rasping, whistling and screaming. There were the rustling of leaves, the croaking of frogs, the hissing of steam, and the flapping of birds' wings. There were clicks from telegraph wires, scraps of talk from other telephones, and curious little squeals that were unlike any known sound. The lines running east and west were noisier than the lines running north and south. The night was noisier than the day, and at the ghostly hour of midnight, for what strange reason no one knows, the babel was at its height. Watson, who had a fanciful mind, suggested that perhaps these sounds were signals from the inhabitants of Mars or some other sociable planet. But the matter-of-fact young telephonists agreed to lay the blame on "induction" - a hazy word which usually meant the natural meddlesomeness of electricity.

Whatever else the mysterious noises were, they were a nuisance. The poor little telephone business was plagued almost out of its senses. It was like a dog with a tin can tied to its tail. No matter where it went, it was pursued by this unearthly clatter. "We were ashamed to present our bills," said A. A. Adee, one of the first agents; "for no matter how plainly a man talked into his telephone, his language was apt to sound like Choctaw at the other end of the line."

All manner of devices were solemnly tried to hush the wires, and each one usually proved to be as futile as an incantation. What was to be done? Step by step the telephone men were driven back. They were beaten. There was no way to silence these noises. Reluctantly, they agreed that the only way was to pull up the ends of each wire from the tainted earth, and join them by a second wire. This was the "metallic circuit" idea. It meant an appalling increase in the use of wire. It would compel the rebuilding of the switchboards and the invention of new signal systems. But it was inevitable; and in 1883, while the dispute about it was in full blast, one of the young men quietly slipped it into use on a new line between Boston and Providence. The effect was magical. "At last," said the delighted manager, "we have a perfectly quiet line."

This young man, a small, slim youth who was twenty-two years old and looked younger, was no other than J. J. Carty, now the first of telephone engineers and almost the creator of his profession. Three years earlier he had timidly asked for a job as operator in the Boston exchange, at five dollars a week, and had shown such an aptitude for the work that he was soon made one of the captains. At thirty years of age he became a central figure in the development of the art of telephony.

What Carty has done is known by telephone men in all

countries; but the story of Carty himself - who he is, and why - is new. First of all, he is Irish, pure Irish. His father had left Ireland as a boy in 1825. During the Civil War his father made guns in the city of Cambridge, where young John Joseph was born; and afterwards he made bells for church steeples. He was instinctively a mechanic and proud of his calling. He could tell the weight of a bell from the sound of it. Moses G. Farmer, the electrical inventor, and Howe, the creator of the sewing-machine, were his friends.

At five years of age, little John J. Carty was taken by his father to the shop where the bells were made, and he was profoundly impressed by the magical strength of a big magnet, that picked up heavy weights as though they were feathers. At the high school his favorite study was physics; and for a time he and another boy named Rolfe - now a distinguished man of science - carried on electrical experiments of their own in the cellar of the Rolfe house. Here they had a "Tom Thumb" telegraph, a telephone which they had ventured to improve, and a hopeless tangle of wires. Whenever they could afford to buy more wires and batteries, they went to a near-by store which supplied electrical apparatus to the professors and students of Harvard. This store, with its workshop in the rear, seemed to the two boys a veritable wonderland; and when Carty, a youth of eighteen, was compelled to leave school because of his bad eyesight, he ran at once and secured the glorious job of being boy-of-all-work in this store of wonders. So, when he became an operator in the Boston telephone exchange, a year later, he had already developed to a remarkable degree his natural genius for telephony.

Since then, Carty and the telephone business have

grown up together, he always a little distance in advance. No other man has touched the apparatus of telephony at so many points. He fought down the flimsy, clumsy methods, which led from one snarl to another. He found out how to do with wires what Dickens did with words. "Let us do it right, boys, and then we won't have any bad dreams" - this has been his motif. And, as the crown and climax of his work, he mapped out the profession of telephone engineering on the widest and most comprehensive lines.

In Carty, the engineer evolved into the educator. His end of the American Telephone and Telegraph Company became the University of the Telephone. He was himself a student by disposition, with a special taste for the writings of Faraday, the forerunner; Tyndall, the expounder; and Spencer, the philosopher. And in 1890, he gathered around him a winnowed group of college graduates - he has sixty of them on his staff to-day - so that he might bequeath to the telephone an engineering corps of loyal and efficient men.

The next problem that faced the young men of the telephone, as soon as they had escaped from the clamor of the mysterious noises, was the necessity of taking down the wires in the city streets and putting them underground. At first, they had strung the wires on poles and roof-tops. They had done this, not because it was cheap, but because it was the only possible way, so far as any one knew in that kindergarten period. A telephone wire required the daintiest of handling. To bury it was to smother it, to make it dull or perhaps entirely useless. But now that the number of wires had swollen from hundreds to thousands, the overhead method had been outgrown.

Some streets in the larger cities had become black with wires. Poles had risen to fifty feet in height, then sixty - seventy - eighty. Finally the highest of all pole lines was built along West Street, New York - every pole a towering Norway pine, with its top ninety feet above the roadway, and carrying thirty cross-arms and three hundred wires.

From poles the wires soon overflowed to housetops, until in New York alone they had overspread eleven thousand roofs. These roofs had to be kept in repair, and their chimneys were the deadly enemies of the iron wires. Many a wire, in less than two or three years, was withered to the merest shred of rust. As if these troubles were not enough, there were the storms of winter, which might wipe out a year's revenue in a single day. The sleet storms were the worst. Wires were weighted down with ice, often three pounds of ice per foot of wire. And so, what with sleet, and corrosion, and the cost of roof-repairing, and the lack of room for more wires, the telephone men were between the devil and the deep sea - between the urgent necessity of burying their wires, and the inexorable fact that they did not know how to do it.

Fortunately, by the time that this problem arrived, the telephone business was fairly well established. It had outgrown its early days of ridicule and incredulity. It was paying wages and salaries and even dividends. Evidently it had arrived on the scene in the nick of time - after the telegraph and before the trolleys and electric lights. Had it been born ten years later, it might not have been able to survive. So delicate a thing as a baby telephone could scarcely have protected itself against the powerful currents of electricity that came into general use in 1886, if it had not first found out a way of hiding safely underground.

The first declaration in favor of an underground system was made by the Boston company in 1880. "It may be expedient to place our entire system underground," said the sorely perplexed manager, "whenever a practicable method is found of accomplishing: it." All manner of theories were afloat but Theodore N. Vail, who was usually the man of constructive imagination in emergencies, began in 1882 a series of actual experiments at Attleborough, Massachusetts, to find out exactly what could, and what could not, be done with wires that were buried in the earth.

A five-mile trench was dug beside a railway track. The work was done handily and cheaply by the labor-saving plan of hitching a locomotive to a plough. Five ploughs were jerked apart before the work was finished. Then, into this trench were laid wires with every known sort of covering. Most of them, naturally, were wrapped with rubber or gutta-percha, after the fashion of a submarine cable. When all were in place, the willing locomotive was harnessed to a huge wooden drag, which threw the ploughed soil back into the trench and covered the wires a foot deep. It was the most professional cable-laying that any one at that time could do, and it succeeded, not brilliantly, but well enough to encourage the telephone engineers to go ahead.

Several weeks later, the first two cables for actual use were laid in Boston and Brooklyn; and in 1883 Engineer J. P. Davis was set to grapple with the Herculean labor of putting a complete underground system in the wire-bound city of New York. This he did in spite of a bombardment of explosions from leaky gas-pipes, and with a woeful lack of experts and standard materials. All manner of makeshifts had to be tried in place of tile ducts, which were not known in 1883. Iron pipe was used at first, then asphalt, concrete, boxes of sand and creosoted wood. As for the wires, they were first wrapped in cotton, and then twisted into cables, usually of a hundred wires each. And to prevent the least taint of moisture, which means sudden death to a telephone current, these cables were invariably soaked in oil.

This oil-filled type of cable carried the telephone business safely through half a dozen years. But it was not the final type. It was preliminary only, the best that could be made at that time. Not one is in use to-day. In 1888 Theodore Vail set on foot a second series of experiments, to see if a cable could be made that was better suited as a highway for the delicate electric currents of the telephone. A young engineer named John A. Barrett, who had already made his mark as an expert, by finding a way to twist and transpose the wires, was set apart to tackle this problem. Being an economical Vermonter, Barrett went to work in a little wooden shed in the backyard of a Brooklyn foundry. In this foundry he had seen a unique machine that could be made to mould hot lead around a rope of twisted wires. This was a notable discovery. It meant TIGHT COVERINGS. It meant a victory over that most troublesome of enemies - moisture. Also, it meant that cables could henceforth be made longer, with fewer sleeves and splices, and without the oil, which had always been an unmitigated nuisance.

Next, having made the cable tight, Barrett set out to produce it more cheaply and by accident stumbled upon a way to make it immensely more efficient. All wires were at that time wrapped with cotton, and his plan was to find some less costly material that would serve the same purpose. One of his workmen, a Virginian, suggested the use of paper twine, which had been used in the South during the Civil War, when cotton was scarce and expensive. Barrett at once searched the South for paper twine and found it. He bought a barrel of it from a small factory in Richmond, but after a trial it proved to be too flimsy. If such paper could be put on flat, he reasoned, it would be stronger. Just then he heard of an erratic genius who had an invention for winding paper tape on wire for the use of milliners.

Paper-wound bonnet-wire! Who could imagine any connection between this and the telephone? Yet this hint was exactly what Barrett needed. He experimented until he had devised a machine that crumpled the paper around the wire, instead of winding it tightly. This was the finishing touch. For a time these paper-wound cables were soaked in oil, but in 1890 Engineer F. A. Pickernell dared to trust to the tightness of the lead sheathing, and laid a "dry core" cable, the first of the modern type, in one of the streets of Philadelphia. This cable was the event of the year. It was not only cheaper. It was the best-talking cable that had ever been harnessed to a telephone.

What Barrett had done was soon made clear. By wrapping the wire with loose paper, he had in reality cushioned it with AIR, which is the best possible insulator. Not the paper, but the air in the paper, had improved the cable. More air was added by the omission of the oil. And presently Barrett perceived that he had merely reproduced in a cable, as far as possible, the conditions of the overhead wires, which are separated by nothing but air.

By 1896 there were two hundred thousand miles of wire snugly wrapped in paper and lying in leaden caskets beneath the streets of the cities, and to-day there are six million miles of it owned by the affiliated Bell companies. Instead of blackening the streets, the wire nerves of the telephone are now out of sight under the roadway, and twining into the basements of buildings like a new sort of metallic ivy. Some cables are so large that a single spool of cable will weigh twenty-six tons and require a giant truck and a sixteen-horse team to haul it to its resting-place. As many as twelve hundred wires are often bunched into one sheath, and each cable lies loosely in a little duct of its own. It is reached by manholes where it runs under the streets and in little switching-boxes placed at intervals it is frayed out into separate pairs of wires that blossom at length into telephones.

Out in the open country there are still the open wires, which in point of talking are the best. In the suburbs of cities there are neat green posts with a single gray cable hung from a heavy wire. Usually, a telephone pole is made from a sixty-year-old tree, a cedar, chestnut, or juniper. It lasts twelve years only, so that the one item of poles is still costing the telephone companies several millions a year. The total number of poles now in the United States, used by telephone and telegraph companies, once covered an area, before they were cut down, as large as the State of Rhode Island.

But the highest triumph of wire-laying came when New York swept into the Skyscraper Age, and when hundreds of tall buildings, as high as the fall of the waters of Niagara, grew up like a range of magical cliffs upon the precious rock of Manhattan. Here the work of the telephone engineer has been so well done that although every room in these cliff-buildings has its telephone, there is not a pole in sight, not a cross-arm, not a wire. Nothing but the tip-ends of an immense system are visible. No sooner is a new skyscraper walled and roofed, than the telephones are in place, at once putting the tenants in touch with the rest of the city and the greater part of the United States. In a single one of these monstrous buildings, the Hudson Terminal, there is a cable that runs from basement to roof and ravels out to reach three thousand desks. This mighty geyser of wires is fifty tons in weight and would, if straightened out into a single line, connect New York with Chicago. Yet it is as invisible as the nerves and muscles of a human body.

During this evolution of the cable, even the wire itself was being remade. Vail and others had noticed that of all the varieties of wire that were for sale, not one was exactly suitable for a telephone system. The first telephone wire was of galvanized iron, which had at least the primitive virtue of being cheap. Then came steel wire, stronger but less durable. But these wires were noisy and not good conductors of electricity. An ideal telephone wire, they found, must be made of either silver or copper. Silver was out of the question, and copper wire was too soft and weak. It would not carry its own weight.

The problem, therefore, was either to make steel wire a better conductor, or to produce a copper wire that would be strong enough. Vail chose the latter, and forthwith gave orders to a Bridgeport manufacturer to begin experiments. A young expert named Thomas B. Doolittle was at once set to work, and presently appeared the first hard-drawn copper wire, made tough-skinned by a fairly simple process. Vail bought thirty pounds of it and scattered it in various parts of the United States, to note the effect upon it of different climates. One length of it may still be seen at the Vail homestead in Lyndonville, Vermont. Then this hard-drawn wire was put to a severe test by being strung between Boston and New York. This line was a brilliant success, and the new wire was hailed with great delight as the ideal servant of the telephone.

Since then there has been little trouble with copper wire, except its price. It was four times as good as iron wire, and four times as expensive. Every mile of it, doubled, weighed two hundred pounds and cost thirty dollars. On the long lines, where it had to be as thick as a lead pencil, the expense seemed to be ruinously great. When the first pair of wires was strung between New York and Chicago, for instance, it was found to weigh 870,000 pounds - a full load for a twenty-two-car freight train; and the cost of the bare metal was $130,000. So enormous has been the use of copper wire since then by the telephone companies, that fully one-fourth of all the capital invested in the telephone has gone to the owners of the copper mines.

For several years the brains of the telephone men were focussed upon this problem - how to reduce the expenditure on copper. One uncanny device, which would seem to be a mere inventor's fantasy if it had not already saved the telephone companies four million dollars or more, is known as the "phantom circuit." It enables three messages to run at the same time, where only two ran before. A double track of wires is made to carry three talk-trains running abreast, a feat made possible by the whimsical disposition of electricity, and which is utterly inconceivable in railroading. This invention, which is the nearest approach as yet to multiple telephony, was conceived by Jacobs in England and Carty in the United States.

But the most copper money has been saved - literally tens of millions of dollars - by persuading thin wires to work as efficiently as thick ones. This has been done by making better transmitters, by insulating the smaller wires with enamel instead of silk, and by placing coils of a certain nature at intervals upon the wires. The invention of this last device startled the telephone men like a flash of lightning out of a blue sky. It came from outside - from the quiet laboratory of a Columbia professor who had arrived in the United States as a young Hungarian immigrant not many years earlier. From this professor, Michael J. Pupin, came the idea of "loading" a telephone line, in such a way as to reinforce the electric current. It enabled a thin wire to carry as far as a thick one, and thus saved as much as forty dollars a wire per mile. As a reward for his cleverness, a shower of gold fell upon Pupin, and made him in an instant as rich as one of the grand-dukes of his native land.

It is now a most highly skilled occupation, supporting fully fifteen thousand families, to put the telephone wires in place and protect them against innumerable dangers. This is the profession of the wire chiefs and their men, a corps of human spiders, endlessly spinning threads under streets and above green fields, on the beds of rivers and the slopes of mountains, massing them in cities and fluffing them out among farms and villages. To tell the doings of a wire chief, in the course of his ordinary week's work, would in itself make a lively book of adventures. Even a washerwoman, with one lone, non-electrical clothes-line of a hundred yards to operate, has often enough trouble with it. But the wire chiefs of the Bell telephone have charge of as much wire as would make TWO HUNDRED MILLION CLOTHES-LINES – ten apiece to every family in the United States; and these lines are not punctuated with clothespins, but with the most delicate of electrical instruments.

The wire chiefs must detect trouble under a thousand disguises. Perhaps a small boy has thrown a snake across the wires or driven a nail into a cable. Perhaps some self-reliant citizen has moved his own telephone from one room to another. Perhaps a sudden rainstorm has splashed its fatal moisture upon an unwiped joint. Or perhaps a submarine cable has been sat upon by the Lusitania and flattened to death. But no matter what the trouble, a telephone system cannot be stopped for repairs. It cannot be picked up and put into a dry-dock. It must be repaired or improved by a sort of vivisection while it is working. It is an interlocking unit, a living, conscious being, half human and half machine; and an injury in any one place may cause a pain or sickness to its whole vast body.

And just as the particles of a human body change every six or seven years, without disturbing the body, so the particles of our telephone systems have changed repeatedly without any interruption of traffic. The constant flood of new inventions has necessitated several complete rebuildings. Little or nothing has ever been allowed to wear out. The New York system was rebuilt three times in sixteen years; and many a costly switchboard has gone to the scrap-heap at three or four years of age. What with repairs and inventions and new construction, the various Bell companies have spent at least $425,000,000 in the first ten years of the twentieth century, without hindering for a day the ceaseless torrent of electrical conversation.

The crowning glory of a telephone system of to-day is not so much the simple telephone itself, nor the maze and mileage of its cables, but rather the wonderful mechanism of the Switchboard. This is the part that will always remain mysterious to the public. It is seldom seen, and it remains as great a mystery to those who have seen it as to those who have not. Explanations of it are futile. As well might any one expect to learn Sanscrit in half an hour as to understand a switchboard by making a tour of investigation around it. It is not like anything else that either man or Nature has ever made. It defies all metaphors and comparisons. It cannot be shown by photography, not even in moving-pictures, because so much of it is concealed inside its wooden body. And few people, if any, are initiated into its inner mysteries except those who belong to its own cortege of inventors and attendants.

A telephone switchboard is a pyramid of inventions. If it is full-grown, it may have two million parts. It may be lit with fifteen thousand tiny electric lamps and nerved with as much wire as would reach from New York to Berlin. It may cost as much as a thousand pianos or as much as three square miles of farms in Indiana. The ten thousand wire hairs of its head are not only numbered, but enswathed in silk, and combed out in so marvellous a way that any one of them can in a flash be linked to any other. Such hair-dressing! Such puffs and braids and ringlet relays! Whoever would learn the utmost that may be done with copper hairs of Titian red, must study the fantastic coiffure of a telephone Switchboard.

If there were no switchboard, there would still be telephones, but not a telephone system. To connect five thousand people by telephone requires five thousand wires when the wires run to a switchboard; but without a switchboard there would have to be 12,497,500 wires - 4,999 to every telephone. As well might there be a nerve-system without a brain, as a telephone system without a switchboard. If there had been at first two separate companies, one owning the telephone and the other the switchboard, neither could have done the business.

Several years before the telephone got a switchboard of its own, it made use of the boards that had been designed for the telegraph. These were as simple as wheelbarrows, and became absurdly inadequate as soon as the telephone business began to grow. Then there came adaptations by the dozen. Every telephone manager became by compulsion an inventor. There was no source of information and each exchange did the best it could. Hundreds of patents were taken out. And by 1884 there had come to be a fairly definite idea of what a telephone switchboard ought to be.

The one man who did most to create the switchboard, who has been its devotee for more than thirty years, is a certain modest and little known inventor, still alive and busy, named Charles E. Scribner. Of the nine thousand switchboard patents, Scribner holds six hundred or more. Ever since 1878, when he devised the first "jackknife switch," Scribner has been the wizard of the switchboard. It was he who saw most clearly its requirements. Hundreds of others have helped, but Scribner was the one man who persevered, who never asked for an easier job, and who in the end became the master of his craft.

It may go far to explain the peculiar genius of Scribner to say that he was born in 1858, in the year of the laying of the Atlantic Cable; and that his mother was at the time profoundly interested in the work and anxious for its success. His father was a judge in Toledo; but young Scribner showed no aptitude for the tangles of the law. He preferred the tangles of wire and system in miniature, which he and several other boys had built and learned to operate. These boys had a benefactor in an old bachelor named Thomas Bond. He had no special interest in telegraphy. He was a dealer in hides. But he was attracted by the cleverness of the boys and gave them money to buy more wires and more batteries. One day he noticed an invention of young Scribner's - a telegraph repeater.

"This may make your fortune," he said, "but no mechanic in Toledo can make a proper model of it for you. You must go to Chicago, where telegraphic apparatus is made." The boy gladly took his advice and went to the Western Electric factory in Chicago. Here he accidentally met Enos M. Barton, the head of the factory. Barton noted that the boy was a genius and offered him a job, which he accepted and has held ever since. Such is the story of the entrance of Charles E. Scribner into the telephone business, where he has been well-nigh indispensable.

His monumental work has been the development of the MULTIPLE Switchboard, a much more brain-twisting problem than the building of the Pyramids or the digging of the Panama Canal. The earlier types of switchboard had become too cumbersome by 1885.

They were well enough for five hundred wires but not for five thousand. In some exchanges as many as half a dozen operators were necessary to handle a single call; and the clamor and confusion were becoming unbearable. Some handier and quieter way had to be devised, and thus arose the Multiple board. The first crude idea of such a way had sprung to life in the brain of a Chicago man named L. B. Firman, in 1879; but he became a farmer and forsook his invention in its infancy.

In the Multiple board, as it grew up under the hands of Scribner, the outgoing wires are duplicated so as to be within reach of every operator. A local call can thus be answered at once by the operator who receives it; and any operator who is overwhelmed by a sudden rush of business can be helped by her companions. Every wire that comes into the board is tasselled out into many ends, and by means of a "busy test," invented by Scribner, only one of these ends can be put into use at a time. The normal limit of such a board is ten thousand wires, and will always remain so, unless a race of long-armed giantesses should appear, who would be able to reach over a greater expanse of board. At present, a business of more than ten thousand lines means a second exchange.

The Multiple board was enormously expensive. It grew more and more elaborate until it cost one-third of a million dollars. The telephone men racked their brains to produce something cheaper to take its place, and they failed. The Multiple boards swallowed up capital as a desert swallows water, but THEY SAVED TEN SECONDS ON EVERY CALL. This was an unanswerable argument in their favor, and by 1887 twenty-one of them were in use.

Since then, the switchboard has had three or four rebuildings. There has seemed to be no limit to the demands of the public or the fertility of Scribner's brain. Persistent changes were made in the system of signalling. The first signal, used by Bell and Watson, was a tap on the diaphragm with the finger-nail. Soon after-wards came a "buzzer," and then the magneto-electric bell. In 1887 Joseph O'Connell, of Chicago, conceived of the use of tiny electric lights as signals, a brilliant idea, as an electric light makes no noise and can be seen either by night or by day. In 1901, J. J. Carty invented the "bridging bell," a way to put four houses on a single wire, with a different signal for each house. This idea made the "party line" practicable, and at once created a boom in the use of the telephone by enterprising farmers.

In 1896 there came a most revolutionary change in switchboards. All things were made new. Instead of individual batteries, one at each telephone, a large common battery was installed in the exchange itself. This meant better signalling and better talking. It reduced the cost of batteries and put them in charge of experts. It established uniformity. It introduced the federal idea into the mechanism of a telephone system. Best of all, it saved FOUR SECONDS ON EVERY CALL. The first of these centralizing switchboards was put in place at Philadelphia; and other cities followed suit as fast as they could afford the expense of rebuilding. Since then, there have come some switchboards that are wholly automatic. Few of these have been put into use, for the reason that a switchboard, like a human body, must be semi-automatic only. To give the most efficient service, there will always need to be an expert to stand between it and the public.

As the final result of all these varying changes in switchboards and signals and batteries, there grew up the modern Telephone Exchange. This is the solar plexus of the telephone body. It is the vital spot. It is the home of the switchboard. It is not any one's invention, as the telephone was. It is a growing mechanism that is not yet finished, and may never be; but it has already evolved far enough to be one of the wonders of the electrical world. There is probably no other part of an American city's equipment that is as sensitive and efficient as a telephone exchange.

The idea of the exchange is somewhat older than the idea of the telephone itself. There were communication exchanges before the invention of the telephone. Thomas B. Doolittle had one in Bridgeport, using telegraph instruments Thomas B. A. David had one in Pittsburg, using printing-telegraph machines, which required little skill to operate. And William A. Childs had a third, for lawyers only, in New York, which used dials at first and afterwards printing machines. These little exchanges had set out to do the work that is done to-day by the telephone, and they did it after a fashion, in a most crude and expensive way. They helped to prepare the way for the telephone, by building up small constituencies that were ready for the telephone when it arrived.

Bell himself was perhaps the first to see the future of the telephone exchange. In a letter written to some English capitalists in 1878, he said: "It is possible to connect every man's house, office or factory with a central station, so as to give him direct communication with his neighbors. . . . It is conceivable that cables of telephone wires could be laid underground, or suspended overhead, connecting by branch wires with private dwellings, shops, etc., and uniting them through the main cable with a central office." This remarkable prophecy has now become stale reading, as stale as Darwin's "Origin of Species," or Adam Smith's "Wealth of Nations." But at the time that it was written it was a most fanciful dream.

When the first infant exchange for telephone service was born in Boston, in 1877, it was the tiny offspring of a burglar-alarm business operated by E. T. Holmes, a young man whose father had originated the idea of protecting property by electric wires in 1858. Holmes was the first practical man who dared to offer telephone service for sale. He had obtained two telephones, numbers six and seven, the first five having gone to the junk-heap; and he attached these to a wire in his burglar-alarm office. For two weeks his business friends played with the telephones, like boys with a fascinating toy; then Holmes nailed up a new shelf in his office, and on this shelf placed six box-telephones in a row. These could be switched into connection with the burglar-alarm wires and any two of the six wires could be joined by a wire cord. Nothing could have been simpler, but it was the arrival of a new idea in the business world.

The Holmes exchange was on the top floor of a little building, and in almost every other city the first exchange was as near the roof as possible, partly to save rent and partly because most of the wires were strung on roof-tops. As the telephone itself had been born in a cellar, so the exchange was born in a garret. Usually, too, each exchange was an off-shoot of some other wire-using business. It was a medley of makeshifts. Almost every part of its outfit had been made for other uses. In Chicago all calls came in to one boy, who bawled them up a speaking-tube to the operators. In another city a boy received the calls, wrote them on white alleys, and rolled them to the boys at the switchboard. There was no number system. Every one was called by name. Even as late as 1880, when New York boasted fifteen hundred telephones, names were still in use. And as the first telephones were used both as transmitters and receivers, there was usually posted up a rule that was highly important: "Don't Talk with your Ear or Listen with your Mouth."

To describe one of those early telephone exchanges in the silence of a printed page is a wholly impossible thing. Nothing but a language of noise could convey the proper impression. An editor who visited the Chicago exchange in 1879 said of it: "The racket is almost deafening. Boys are rushing madly hither and thither, while others are putting in or taking out pegs from a central framework as if they were lunatics engaged in a game of fox and geese." In the same year E. J. Hall wrote from Buffalo that his exchange with twelve boys had become "a perfect Bedlam." By the clumsy methods of those days, from two to six boys were needed to handle each call. And as there was usually more or less of a cat-and-dog squabble between the boys and the public, with every one yelling at the top of his voice, it may be imagined that a telephone exchange was a loud and frantic place.

Boys, as operators, proved to be most complete and consistent failures. Their sins of omission and commission would fill a book. What with whittling the switchboards, swearing at subscribers, playing tricks with the wires, and roaring on all occasions like young bulls of Bashan, the boys in the first exchanges did their full share in adding to the troubles of the business. Nothing could be done with them. They were immune to all schemes of discipline. Like the MYSTERIOUS NOISES they could not be controlled, and by general consent they were abolished. In place of the noisy and obstreperous boy came the docile, soft-voiced girl.

If ever the rush of women into the business world was an unmixed blessing, it was when the boys of the telephone exchanges were superseded by girls. Here at its best was shown the influence of the feminine touch. The quiet voice, pitched high, the deft fingers, the patient courtesy and attentiveness - these qualities were precisely what the gentle telephone required in its attendants. Girls were easier to train; they did not waste time in retaliatory conversation; they were more careful; and they were much more likely to give "the soft answer that turneth away wrath."

A telephone call under the boy regime meant Bedlam and five minutes; afterwards, under the girl regime, it meant silence and twenty seconds. Instead of the incessant tangle and tumult, there came a new species of exchange - a quiet, tense place, in which several score of young ladies sit and answer the language of the switchboard lights. Now and then, not often, the signal lamps flash too quickly for these expert phonists. During the panic of 1907 there was one mad hour when almost every telephone in Wall Street region was being rung up by some desperate speculator. The switchboards were ablaze with lights. A few girls lost their heads. One fainted and was carried to the rest-room. But the others flung the flying shuttles of talk until, in a single exchange fifteen thousand conversations had been made possible in sixty minutes. There are always girls in reserve for such explosive occasions, and when the hands of any operator are seen to tremble, and she has a warning red spot on each cheek, she is taken off and given a recess until she recovers her poise.

These telephone girls are the human part of a great communication machine. They are weaving a web of talk that changes into a new pattern every minute. How many possible combinations there are with the five million telephones of the Bell System, or what unthinkable mileage of conversation, no one has ever dared to guess. But whoever has once seen the long line of white arms waving back and forth in front of the switchboard lights must feel that he has looked upon the very pulse of the city's life.

In 1902 the New York Telephone Company started a school, the first of its kind in the world, for the education of these telephone girls. This school is hidden amid ranges of skyscrapers, but seventeen thousand girls discover it in the course of the year. It is a most particular and exclusive school. It accepts fewer than two thousand of these girls, and rejects over fifteen thousand. Not more than one girl in every eight can measure up to its standards; and it cheerfully refuses as many students in a year as would make three Yales or Harvards.

This school is unique, too, in the fact that it charges no fees, pays every student five dollars a week, and then provides her with a job when she graduates. But it demands that every girl shall be in good health, quick-handed, clear-voiced, and with a certain poise and alertness of manner. Presence of mind, which, in Herbert Spencer's opinion, ought to be taught in every university, is in various ways drilled into the temperament of the telephone girl. She is also taught the knack of concentration, so that she may carry the switchboard situation in her head, as a chess-player carries in his head the arrangement of the chess-men. And she is much more welcome at this strange school if she is young and has never worked in other trades, where less speed and vigilance are required.

No matter how many millions of dollars may be spent upon cables and switchboards, the quality of telephone service depends upon the girl at the exchange end of the wire. It is she who meets the public at every point. She is the despatcher of all the talk trains; she is the ruler of the wire highways; and she is expected to give every passenger-voice an instantaneous express to its destination. More is demanded from her than from any other servant of the public. Her clients refuse to stand in line and quietly wait their turn, as they are quite willing to do in stores and theatres and barber shops and railway stations and everywhere else. They do not see her at work and they do not know what her work is. They do not notice that she answers a call in an average time of three and a half seconds. They are in a hurry, or they would not be at the telephone; and each second is a minute long. Any delay is a direct personal affront that makes a vivid impression upon their minds. And they are not apt to remember that most of the delays and blunders are being made, not by the expert girls, but by the careless people who persist in calling wrong numbers and in ignoring the niceties of telephone etiquette.

The truth about the American telephone girl is that she has become so highly efficient that we now expect her to be a paragon of perfection. To give the young lady her due, we must acknowledge that she has done more than any other person to introduce courtesy into the business world. She has done most to abolish the old-time roughness and vulgarity. She has made big business to run more smoothly than little business did, half a century ago. She has shown us how to take the friction out of conversation, and taught us refinements of politeness which were rare even among the Beau Brummels of pre-telephonic days. Who, for instance, until the arrival of the telephone girl, appreciated the difference between "Who are you?" and "Who is this?" Or who else has so impressed upon us the value of the rising inflection, as a gentler habit of speech? This propaganda of politeness has gone so far that to-day the man who is profane or abusive at the telephone, is cut off from the use of it. He is cast out as unfit for a telephone-using community.

And now, so that there shall be no anticlimax in this story of telephone development, we must turn the spotlight upon that immense aggregation of workshops in which have been made three-fifths of the telephone apparatus of the world - the Western Electric. The mother factory of this globe-trotting business is the biggest thing in the spacious back-yard of Chicago, and there are eleven smaller factories - her children - scattered over the earth from New York to Tokio. To put its totals into a sentence, it is an enterprise of 26,000-man-power, and 40,000,000-dollar-power; and the telephonic goods that it produces in half a day are worth one hundred thousand dollars - as much, by the way, as the Western Union REFUSED to pay for the Bell patents in 1877.

The Western Electric was born in Chicago, in the ashes of the big fire of 1871; and it has grown up to its present greatness quietly, without celebrating its birthdays. At first it had no telephones to make. None had been invented, so it made telegraphic apparatus, burglar-alarms, electric pens, and other such things. But in 1878, when the Western Union made its short-lived attempt to compete with the Bell Company, the Western Electric agreed to make its telephones. Three years later, when the brief spasm of competition was ended, the Western Electric was taken in hand by the Bell people and has since then remained the great workshop of the telephone.

The main plant in Chicago is not especially remarkable from a manufacturing point of view. Here are the inevitable lumber-yards and foundries and machine-shops. Here is the mad waltz of the spindles that whirl silk and cotton threads around the copper wires, very similar to what may be seen in any braid factory. Here electric lamps are made, five thousand of them in a day, in the same manner as elsewhere, except that here they are so small and dainty as to seem designed for fairy palaces,

The things that are done with wire in the Western Electric factories are too many for any mere outsider to remember. Some wire is wrapped with paper tape at a speed of nine thousand miles a day. Some is fashioned into fantastic shapes that look like absurd sea-monsters, but which in reality are only the nerve systems of switchboards. And some is twisted into cables by means of a dozen whirling drums - a dizzying sight, as each pair of drums revolve in opposite directions. Because of the fact that a cable's inevitable enemy is moisture, each cable is wound on an immense spool and rolled into an oven until it is as dry as a cinder. Then it is put into a strait-jacket of lead pipe, sealed at both ends, and trundled into a waiting freight car.

No other company uses so much wire and hard rubber, or so many tons of brass rods, as the Western Electric. Of platinum, too, which is more expensive than gold, it uses one thousand pounds a year in the making of telephone transmitters. This is imported from the Ural Mountains. The silk thread comes from Italy and Japan; the iron for magnets, from Norway; the paper tape, from Manila; the mahogany, from South America; and the rubber, from Brazil and the valley of the Congo. At least seven countries must cooperate to make a telephone message possible.

Perhaps the most extraordinary feature in the Western Electric factories is the multitude of its inspectors. No other sort of manufacturing, not even a Government navy-yard, has so many. Nothing is too small to escape these sleuths of inspection. They test every tiny disc of mica, and throw away nine out of ten. They test every telephone by actual talk, set up every switchboard, and try out every cable. A single transmitter, by the time it is completed, has had to pass three hundred examinations; and a single coin-box is obliged to count ten thousand nickels before it graduates into the outer world. Seven hundred inspectors are on guard in the two main plants at Chicago and New York. This is a ruinously large number, from a profit-making point of view; but the inexorable fact is that in a telephone system nothing is insignificant. It is built on such altruistic lines that an injury to any one part is the concern of all.

As usual, when we probe into the history of a business that has grown great and overspread the earth, we find a Man; and the Western Electric is no exception to this rule. Its Man, still fairly hale and busy after forty years of leadership, is Enos M. Barton. His career is the typical American story of self-help. He was a telegraph messenger boy in New York during the Civil War, then a telegraph operator in Cleveland. In 1869 his salary was cut down from one hundred dollars a month to ninety dollars; whereupon he walked out and founded the Western Electric in a shabby little machine-shop. Later he moved to Chicago, took in Elisha Gray as his partner, and built up a trade in the making of telegraphic materials.

When the telephone was invented, Barton was one of the sceptics. "I well remember my disgust," he said, "when some one told me it was possible to send conversation along a wire." Several months later he saw a telephone and at once became one of its apostles. By 1882 his plant had become the official workshop of the Bell Companies. It was the headquarters of invention and manufacturing. Here was gathered a notable group of young men, brilliant and adventurous, who dared to stake their futures on the success of the telephone. And always at their head was Barton, as a sort of human switchboard, who linked them all together and kept them busy.

In appearance, Enos M. Barton closely resembles ex-President Eliot, of Harvard. He is slow in speech, simple in manner, and with a rare sagacity in business affairs. He was not an organizer, in the modern sense. His policy was to pick out a man, put him in a responsible place, and judge him by results. Engineers could become bookkeepers, and bookkeepers could become engineers. Such a plan worked well in the earlier days, when the art of telephony was in the making, and when there was no source of authority on telephonic problems. Barton is the bishop emeritus of the Western Electric to-day; and the big industry is now being run by a group of young hustlers, with H. B. Thayer at the head of the table. Thayer is a Vermonter who has climbed the ladder of experience from its lower rungs to the top. He is a typical Yankee - lean, shrewd, tireless, and with a cold-blooded sense of justice that fits him for the leadership of twenty-six thousand people.

So, as we have seen, the telephone as Bell invented it, was merely a brilliant beginning in the development of the art of telephony. It was an elfin birth - an elusive and delicate sprite that had to be nurtured into maturity. It was like a soul, for which a body had to be created; and no one knew how to make such a body. Had it been born in some less energetic country, it might have remained feeble and undeveloped; but not in the United States. Here in one year it had become famous, and in three years it had become rich. Bell's invincible patent was soon buttressed by hundreds of others. An open-door policy was adopted for invention. Change followed change to such a degree that the experts of 1880 would be lost to-day in the mazes of a telephone exchange.

The art of the telephone engineer has in thirty years grown from the most crude and clumsy of experiments into an exact and comprehensive profession. As Carty has aptly said, "At first we invariably approached every problem from the wrong end. If we had been told to load a herd of cattle on a steamer, our method would have been to hire a Hagenbeck to train the cattle for a couple of years, so that they would know enough to walk aboard of the ship when he gave the signal; but to-day, if we had to ship cattle, we would know enough to make a greased chute and slide them on board in a jiffy."

The telephone world has now its own standards and ideals. It has a language of its own, a telephonese that is quite unintelligible to outsiders. It has as many separate branches of study as medicine or law. There are few men, half a dozen at most, who can now be said to have a general knowledge of telephony. And no matter how wise a telephone expert may be, he can never reach perfection, because of the amazing variety of things that touch or concern his profession.

"No one man knows all the details now," said Theodore Vail. "Several days ago I was walking through a telephone exchange and I saw something new. I asked Mr. Carty to explain it. He is our chief engineer; but he did not understand it. We called the manager. He did n't know, and called his assistant. He did n't know, and called the local engineer, who was able to tell us what it was."

To sum up this development of the art of telephony - to present a bird's-eye view - it may be divided into four periods:

1. Experiment. 1876 to 1886. This was the period of invention, in which there were no experts and no authorities. Telephonic apparatus consisted of makeshifts and adaptations. It was the period of iron wire, imperfect transmitters, grounded circuits, boy operators, peg switchboards, local batteries, and overhead lines.

2. Development. 1886 to 1896. In this period amateurs became engineers. The proper type of apparatus was

discovered, and was improved to a high point of efficiency. In this period came the multiple switchboard, copper wire, girl operators, underground cables, metallic circuit, common battery, and the long-distance lines.

3. Expansion. 1896 to 1906. This was the era of big business. It was an autumn period, in which the telephone men and the public began to reap the fruits of twenty years of investment and hard work. It was the period of the message rate, the pay station, the farm line, and the private branch exchange.

4. Organization. 1906 - . With the success of the Pupin coil, there came a larger life for the telephone. It became less local and more national. It began to link together its scattered parts. It discouraged the waste and anarchy of duplication. It taught its older, but smaller brother, the telegraph, to cooperate. It put itself more closely in touch with the will of the public. And it is now pushing ahead, along the two roads of standardization and efficiency, toward its ideal of one universal telephone system for the whole nation. The key-word of the telephone development of to-day is this - organization.

CHAPTER V

THE EXPANSION OF THE BUSINESS

The telephone business did not really begin to grow big and overspread the earth until 1896, but the keynote of expansion was first sounded by Theodore Vail in the earliest days, when as yet the telephone was a babe in arms. In 1879 Vail said, in a letter written to one of his captains:

"Tell our agents that we have a proposition on foot to connect the different cities for the purpose of personal communication, and in other ways to organize a GRAND TELEPHONIC SYSTEM."

This was brave talk at that time, when there were not in the whole world as many telephones as there are today in Cincinnati. It was brave talk in those days of iron wire, peg switchboards, and noisy diaphragms. Most telephone men regarded it as nothing more than talk. They did not see any business future for the telephone except in short-distance service. But Vail was in earnest. His previous experience as the head of the railway mail service had lifted him up to a higher point of view. He knew the need of a national system of communication that would be quicker and more direct than either the telegraph or the post office.

"I saw that if the telephone could talk one mile to-day," he said, "it would be talking a hundred miles to-morrow." And he persisted, in spite of a considerable deal of ridicule, in maintaining that the telephone was destined to connect cities and nations as well as individuals.

Four months after he had prophesied the "grand telephonic system," he encouraged Charles J. Glidden, of world-tour fame, to build a telephone line between Boston and Lowell. This was the first inter-city line. It was well placed, as the owners of the Lowell mills lived in Boston, and it made a small profit from the start. This success cheered Vail on to a master-effort. He resolved to build a line from Boston to Providence, and was so stubbornly bent upon doing this that when the Bell Company refused to act, he picked up the risk and set off with it alone. He organized a company of well-known Rhode Islanders - nicknamed the "Governors' Company" - and built the line. It was a failure at first, and went by the name of "Vail's Folly." But Engineer Carty, by a happy thought, DOUBLED THE WIRE, and thus in a moment established two new factors in the telephone business - the Metallic Circuit and the Long Distance line.

At once the Bell Company came over to Vail's point of view, bought his new line, and launched out upon what seemed to be the foolhardy enterprise of stringing a double wire from Boston to New York. This was to be not only the longest of all telephone lines, strung on ten thousand poles; it was to be a line de luxe, built of glistening red copper, not iron. Its cost was to be seventy thousand dollars, which was an enormous sum in those hardscrabble days. There was much opposition to such extravagance, and much ridicule. "I would n't take that line as a gift," said one of the Bell Company's officials.

But when the last coil of wire was stretched into place, and the first "Hello" leaped from Boston to New York, the new line was a victorious success. It carried messages from the first day; and more, it raised the whole telephone business to a higher level. It swept away the prejudice that telephone service could become nothing more than a neighborhood affair. "It was the salvation of the business," said Edward J. Hill. It marked a turning-point in the history of the telephone, when the day of small things was ended and the day of great things was begun. No one man, no hundred men, had created it. It was the final result of ten years of invention and improvement.

While this epoch-making line was being strung, Vail was pushing his "grand telephonic system" policy by organizing The American Telephone and Telegraph Company. This, too, was a master-stroke. It was the introduction of the staff-and-line method of organization into business. It was doing for the forty or fifty Bell Companies what Von Moltke did for the German army prior to the Franco-Prussian War. It was the creation of a central company that should link all local companies together, and itself own and operate the means by which these companies are united. This central company was to grapple with all national problems, to own all telephones and long-distance lines, to protect all patents, and to be the headquarters of invention, information, capital, and legal protection for the entire federation of Bell Companies.

Seldom has a company been started with so small a capital and so vast a purpose. It had no more than

$100,000 of capital stock, in 1885; but its declared object was nothing less than to establish a system of wire communication for the human race. Here are, in its own words, the marching orders of this Company: "To connect one or more points in each and every city, town, or place an the State of New York, with one or more points in each and every other city, town, or place in said State, and in each and every other of the United States, and in Canada, and Mexico; and each and every of said cities, towns, and places is to be connected with each and every other city, town, or place in said States and countries, and also by cable and other appropriate means with the rest of the known world."

So ran Vail's dream, and for nine years he worked mightily to make it come true. He remained until the various parts of the business had grown together, and until his plan for a "grand telephonic system" was under way and fairly well understood. Then he went out, into a series of picturesque enterprises, until he had built up a four-square fortune; and recently, in 1907, he came back to be the head of the telephone business, and to complete the work of organization that he started thirty years before.

When Vail said auf wiedersehen to the telephone business, it had passed from infancy to childhood. It was well shaped but not fully grown. Its pioneering days were over. It was self-supporting and had a little money in the bank. But it could not then have carried the load of traffic that it carries to-day. It had still too many problems to solve and too much general inertia to overcome. It needed to be conserved, drilled, educated, popularized. And the man who was finally chosen to replace Vail was in many respects the appropriate leader for such a preparatory period.

Hudson - John Elbridge Hudson - was the name of the new head of the telephone people. He was a man of middle age, born in Lynn and bred in Boston; a long-pedigreed New Englander, whose ancestors had smelted iron ore in Lynn when Charles the First was King. He was a lawyer by profession and a university professor by temperament. His specialty, as a man of affairs, had been marine law; and his hobby was the collection of rare books and old English engravings. He was a master of the Greek language, and very fond of using it. On all possible occasions he used the language of Pericles in his conversation; and even carried this preference so far as to write his business memoranda in Greek. He was above all else a scholar, then a lawyer, and somewhat incidentally the central figure in the telephone world.

But it was of tremendous value to the telephone business at that time to have at its head a man of Hudson's intellectual and moral calibre.

He gave it tone and prestige. He built up its credit. He kept it clean and clear above all suspicion of wrong-doing. He held fast whatever had been gained. And he prepared the way for the period of expansion by borrowing fifty millions for improvements, and by adding greatly to the strength and influence of the American Telephone and Telegraph Company.

Hudson remained at the head of the telephone table until his death, in 1900, and thus lived to see the dawn of the era of big business. Under his regime great things were done in the development of the art. The business was pushed ahead at every point by its captains. Every man in his place, trying to give a little better service than yesterday - that was the keynote of the Hudson period. There was no one preeminent genius. Each important step forward was the result of the cooperation of many minds, and the prodding necessities of a growing traffic.

By 1896, when the Common Battery system created a new era, the telephone engineer had pretty well mastered his simpler troubles. He was able to handle his wires, no matter how many. By this time, too, the public was ready for the telephone. A new generation had grown up, without the prejudices of its fathers. People had grown away from the telegraphic habit of thought, which was that wire communications were expensive luxuries for the few. The telephone was, in fact, a new social nerve, so new and so novel that very nearly twenty years went by before it had fully grown into place, and before the social body developed the instinct of using it.

Not that the difficulties of the telephone engineers were over, for they were not. They have seemed to grow more numerous and complex every year. But by 1896 enough had been done to warrant a forward movement. For the next ten-year period the keynote of telephone history was EXPANSION. Under the prevailing flat-rate plan of payment, all customers paid the same yearly price and then used their telephones as often as they pleased. This was a simple method, and the most satisfactory for small towns and farming regions. But in a great city such a plan grew to be suicidal. In New York, for instance, the price had to be raised to $240, which lifted the telephone as high above the mass of the citizens as though it were a piano or a diamond sunburst. Such a plan was strangling the business. It was shutting out the small users. It was clogging the wires with deadhead calls. It was giving some people too little service and others too much. It was a very unsatisfactory situation.

How to extend the service and at the same time cheapen it to small users - that was the Gordian knot; and the man who unquestionably did most to untie it was Edward J. Hall. Mr. Hall founded the telephone business in Buffalo in 1878, and seven years afterwards became the chief of the long-distance traffic. He was then, and is to-day, one of the statesmen of the telephone. For more than thirty years he has been the "candid friend" of the business, incessantly suggesting, probing, and criticising. Keen and dispassionate, with a genius for mercilessly cutting to the marrow of a proposition, Hall has at the same time been a zealot for the improvement and extension of telephone service. It was he who set the agents free from the ball-and-chain of royalties, allowing them to pay instead a percentage of gross receipts. And it was he who "broke the jam," as a lumberman would say, by suggesting the MESSAGE RATE system.

By this plan, which U. N. Bethell developed to its highest point in New York, a user of the telephone pays a fixed minimum price for a certain number of messages per year, and extra for all messages over this number. The large user pays more, and the little user pays less. It opened up the way to such an expansion of telephone business as Bell, in his rosiest dreams, had never imagined. In three years, after 1896, there were twice as many users; in six years there were four times as many; in ten years there were eight to one. What with the message rate and the pay station, the telephone was now on its way to be universal. It was adapted to all kinds and conditions of men. A great corporation, nerved at every point with telephone wires, may now pay fifty thousand dollars to the Bell Company, while at the same time a young Irish immigrant boy, just arrived in New York City, may offer five coppers and find at his disposal a fifty million dollar telephone system.

When the message rate was fairly well established, Hudson died - fell suddenly to the ground as he was about to step into a railway carriage. In his place came Frederick P. Fish, also a lawyer and a Bostonian. Fish was a popular, optimistic man, with a "full-speed-ahead" temperament. He pushed the policy of expansion until he broke all the records. He borrowed money in stupendous amounts - $150,000,000 at one time - and flung it into a campaign of red-hot development. More business he demanded, and more, and more, until his captains, like a thirty-horse team of galloping horses, became very nearly uncontrollable.

It was a fast and furious period. The whole country was ablaze with a passion of prosperity. After generations of conflict, the men with large ideas had at last put to rout the men of small ideas. The waste and folly of competition had everywhere driven men to the policy of cooperation. Mills were linked to mills and factories to factories, in a vast mutualism of industry such as no other age, perhaps, has ever known. And as the telephone is essentially the instrument of co-working and interdependent people, it found itself suddenly welcomed as the most popular and indispensable of all the agencies that put men in touch with each other.

To describe this growth in a single sentence, we might

say that the Bell telephone secured its first million of capital in 1879; its first million of earnings in 1882; its first million of dividends in 1884; its first million of surplus in 1885. It had paid out its first million for legal expenses by 1886; began first to send a million messages a day in 1888; had strung its first million miles of wire in 1900; and had installed its first million telephones in 1898. By 1897 it had spun as many cobwebs of wire as the mighty Western Union itself; by 1900 it had twice as many miles of wire as the Western Union, and in 1905 FIVE TIMES as many. Such was the plunging progress of the Bell Companies in this period of expansion, that by 1905 they had swept past all European countries combined, not only in the quality of the service but in the actual number of telephones in use. This, too, without a cent of public money, or the protection of a tariff, or the prestige of a governmental bureau.

By 1892 Boston and New York were talking to Chicago, Milwaukee, Pittsburg, and Washington. One-half of the people of the United States were within talking distance of each other. The THOUSAND-MILE TALK had ceased to be a fairy tale. Several years later the western end of the line was pushed over the plains to Nebraska, enabling the spoken word in Boston to be heard in Omaha. Slowly and with much effort the public were taught to substitute the telephone for travel. A special long-distance salon was fitted up in New York City to entice people into the habit of talking to other cities. Cabs were sent for customers; and when one arrived, he was escorted over Oriental rugs to a gilded booth, draped with silken curtains. This was the famous "Room Nine." By such and many other allurements a larger idea of telephone service was given to the public mind; until in 1909 at least eighteen thousand New York-Chicago conversations were held, and the revenue from strictly long-distance messages was twenty-two thousand dollars a day.

By 1906 even the Rocky Mountain Bell Company had grown to be a ten-million-dollar enterprise. It began at Salt Lake City with a hundred telephones, in 1880. Then it reached out to master an area of four hundred and thirteen thousand square miles - a great Lone Land of undeveloped resources. Its linemen groped through dense forests where their poles looked like toothpicks beside the towering pines and cedars. They girdled the mountains and basted the prairies with wire, until the lonely places were brought together and made sociable. They drove off the Indians, who wanted the bright wire for ear-rings and bracelets; and the bears, which mistook the humming of the wires for the buzzing of bees, and persisted in gnawing the poles down. With the most heroic optimism, this Rocky Mountain Company persevered until, in 1906, it had created a seventy-thousand-mile nerve-system for the far West.

Chicago, in this year, had two hundred thousand telephones in use, in her two hundred square miles of area. The business had been built up by General Anson Stager, who was himself wealthy, and able to attract the support of such men as John Crerar, H. H. Porter, and Robert T. Lincoln. Since 1882 it has paid dividends, and in one glorious year its stock soared to four hundred dollars a share. The old-timers - the men who clambered over roof-tops in 1878 and tacked iron wires wherever they could without being chased off - are still for the most part in control of the Chicago company.

But as might have been expected, it was New York City that was the record-breaker when the era of telephone expansion arrived. Here the flood of big business struck with the force of a tidal wave. The number of users leaped from 56,000 in 1900 up to 810,000 in 1908. In a single year of sweating and breathless activity, 65,000 new telephones were put on desks or hung on walls - an average of one new user for every two minutes of the business day.

Literally tons, and hundreds of tons, of telephones were hauled in drays from the factory and put in place in New York's homes and offices. More and more were demanded, until to-day there are more telephones in New York than there are in the four countries, France, Belgium, Holland, and Switzerland combined. As a user of telephones New York has risen to be unapproachable. Mass together all the telephones of London, Glasgow, Liverpool, Manchester, Birmingham, Leeds, Sheffleld, Bristol, and Belfast, and there will even then be barely as many as are carrying the conversations of this one American city.

In 1879 the New York telephone directory was a small card, showing two hundred and fifty-two names; but now it has grown to be an eight-hundred-page quarterly, with a circulation of half a million, and requiring twenty drays, forty horses, and four hundred men to do the work of distribution. There was one shabby little exchange thirty years ago; but now there are fifty-two exchanges, as the nerve-centres of a vast fifty-million-dollar system. Incredible as it may seem to foreigners, it is literally true that in a single building in New York, the Hudson Terminal, there are more telephones than in Odessa or Madrid, more than in the two kingdoms of Greece and Bulgaria combined.

Merely to operate this system requires an army of more than five thousand girls. Merely to keep their records requires two hundred and thirty-five million sheets of paper a year. Merely to do the writing of these records wears away five hundred and sixty thousand lead pencils. And merely to give these girls a cup of tea or coffee at noon, compels the Bell Company to buy yearly six thousand pounds of tea, seventeen thousand pounds of coffee, forty-eight thousand cans of condensed milk, and one hundred and forty barrels of sugar.

The myriad wires of this New York system are tingling with talk every minute of the day and night. They are most at rest between three and four o'clock in the morning, although even then there are usually ten calls a minute. Between five and six o'clock, two thousand New Yorkers are awake and at the telephone. Half an hour later there are twice as many. Between seven and eight twenty-five thousand people have called up twenty-five thousand other people, so that there are as many people talking by wire as there were in the whole city of New York in the Revolutionary period. Even this is only the dawn of the day's business. By half-past eight it is doubled; by nine it is trebled; by ten it is multiplied sixfold; and by eleven the roar has become an incredible babel of one hundred and eighty thousand conversations an hour, with fifty new voices clamoring at the exchanges every second.

This is "the peak of the load." It is the topmost pinnacle of talk. It is the utmost degree of service that the telephone has been required to give in any city. And it is as much a world's wonder, to men and women of imagination, as the steel mills of Homestead or the turbine leviathans that curve across the Atlantic

Ocean in four and a half days.

As to the men who built it up: Charles F. Cutler died in 1907, but most of the others are still alive and busy. Union N. Bethell, now in Cutler's place at the head of the New York Company, has been the operating chief for eighteen years. He is a man of shrewdness and sympathy, with a rare sagacity in solving knotty problems, a president of the new type, who regards his work as a sort of obligation he owes to the public. And just as foreigners go to Pittsburg to see the steel business at its best; just as they go to Iowa and Kansas to see the New Farmer, so they make pilgrimages to Bethell's office to learn the profession of telephony.

This unparalleled telephone system of New York grew up without having at any time the rivalry of competition. But in many other cities and especially in the Middle West, there sprang up in 1895 a medley of independent companies. The time of the original patents had expired, and the Bell Companies found themselves freed from the expense of litigation only to be snarled up in a tangle of duplication. In a few years there were six thousand of these little Robinson Crusoe companies. And by 1901 they had put in use more than a million telephones and were professing to have a capital of a hundred millions.

Most of these companies were necessary and did much to expand the telephone business into new territory. They were in fact small mutual associations of a dozen or a hundred farmers, whose aim was to get telephone service at cost. But there were other companies, probably a thousand or more, which were organized by promoters who built their hopes on the fact that the Bell Companies were unpopular, and on the myth that they were fabulously rich. Instead of legitimately extending telephone lines into communities that had none, these promoters proceeded to inflict the messy snarl of an overlapping system upon whatever cities would give them permission to do so.

In this way, masked as competition, the nuisance and waste of duplication began in most American cities. The telephone business was still so young, it was so little appreciated even by the telephone officials and engineers, that the public regarded a second or a third telephone system in one city as quite a possible and desirable innovation. "We have two ears," said one promoter; "why not therefore have two telephones?"

This duplication went merrily on for years before it was generally discovered that the telephone is not an ear, but a nerve system; and that such an experiment as a duplicate nerve system has never been attempted by Nature, even in her most frivolous moods. Most people fancied that a telephone system was practically the same as a gas or electric light system, which can often be duplicated with the result of cheaper rates and better service. They did not for years discover that two telephone companies in one city means either half service or double cost, just as two fire departments or two post offices would.

Some of these duplicate companies built up a complete plant, and gave good local service, while others proved to be mere stock bubbles. Most of them were over-capitalized, depending upon public sympathy to atone for deficiencies in equipment. One which had printed fifty million dollars of stock for sale was sold at auction in 1909 for four hundred thousand dollars. All told, there were twenty-three of these bubbles that burst in 1905, twenty-one in 1906, and twelve in 1907. So high has been the death-rate among these isolated companies that at a recent convention of telephone agents, the chairman's gavel was made of thirty-five pieces of wood, taken from thirty-five switchboards of thirty-five extinct companies.

A study of twelve single-system cities and twenty-seven double-system cities shows that there are about eleven per cent more telephones under the double-system, and that where the second system is put in, every fifth user is obliged to pay for two telephones. The rates are alike, whether a city has one or two systems. Duplicating companies raised their rates in sixteen cities out of the twenty-seven, and reduced them in one city. Taking the United States as a whole, there are to-day fully two hundred and fifty thousand people who are paying for two telephones instead of one, an economic waste of at least ten million dollars a year.

A fair-minded survey of the entire independent telephone movement would probably show that it was at first a stimulant, followed, as stimulants usually are, by a reaction. It was unquestionably for several years a spur to the Bell Companies. But it did not fulfil its promises of cheap rates, better service, and high dividends; it did little or nothing to improve telephonic apparatus, producing nothing new except the automatic switchboard - a brilliant invention, which is now in its experimental period. In the main, perhaps, it has been a reactionary and troublesome movement in the cities, and a progressive movement among the farmers.

By 1907 it was a wave that had spent its force. It was no longer rolling along easily on the broad ocean of hope, but broken and turned aside by the rocks of actual conditions. One by one the telephone promoters learned the limitations of an isolated company, and asked to be included as members of the Bell family. In 1907 four hundred and fifty-eight thousand independent telephones were linked by wire to the nearest Bell Company; and in 1908 these were followed by three hundred and fifty thousand more. After this landslide to the policy of consolidation, there still remained a fairly large assortment of independent companies; but they had lost their dreams and their illusions.

As might have been expected, the independent movement produced a number of competent local leaders, but none of national importance. The Bell Companies, on the other hand, were officered by men who had for a quarter of a century been surveying telephone problems from a national point of view. At their head, from 1907 onwards, was Theodore N. Vail, who had returned dramatically, at the precise moment when he was needed, to finish the work that he had begun in 1878. He had been absent for twenty years, developing water-power and building street-railways in South America. In the first act of the telephone drama, it was he who put the enterprise upon a business basis, and laid down the first principles of its policy. In the second and third acts he had no place; but when the curtain rose upon the fourth act, Vail was once more the central figure, standing white-haired among his captains, and pushing forward the completion of the "grand telephonic system" that he had dreamed of when the telephone was three years old.

Thus it came about that the telephone business was created by Vail, conserved by Hudson, expanded by Fish, and is now in process of being consolidated by

Vail. It is being knit together into a stupendous Bell System - a federation of self-governing companies, united by a central company that is the busiest of them all. It is no longer protected by any patent monopoly. Whoever is rich enough and rash enough may enter the field. But it has all the immeasurable advantages that come from long experience, immense bulk, the most highly skilled specialists, and an abundance of capital. "The Bell System is strong," says Vail, "because we are all tied up together; and the success of one is therefore the concern of all."

The Bell System! Here we have the motif of American telephone development. Here is the most comprehensive idea that has entered any telephone engineer's brain. Already this Bell System has grown to be so vast, so nearly akin to a national nerve system, that there is nothing else to which we can compare it. It is so wide-spread that few are aware of its greatness. It is strung out over fifty thousand cities and communities.

If it were all gathered together into one place, this Bell System, it would make a city of Telephonia as large as Baltimore. It would contain half of the telephone property of the world. Its actual wealth would be fully $760,000,000, and its revenue would be greater than the revenue of the city of New York.

Part of the property of the city of Telephonia consists of ten million poles, as many as would make a fence from New York to California, or put a stockade around Texas. If the Telephonians wished to use these poles at home, they might drive them in as piles along their water-front, and have a twenty-five thousand-acre dock; or if their city were a hundred square miles in extent, they might set up a seven-ply wall around it with these poles.

Wire, too! Eleven million miles of it! This city of Telephonia would be the capital of an empire of wire. Not all the men in New York State could shoulder this burden of wire and carry it. Throw all the people of Illinois in one end of the scale, and put on the other side the wire-wealth of Telephonia, and long before the last coil was in place, the Illinoisans would be in the air.

What would this city do for a living? It would make two-thirds of the telephones, cables, and switchboards of all countries. Nearly one-quarter of its citizens would work in factories, while the others would be busy in six thousand exchanges, making it possible for the people of the United States to talk to one another at the rate of SEVEN THOUSAND MILLION CONVERSATIONS A YEAR.

The pay-envelope army that moves to work every morning in Telephonia would be a host of one hundred and ten thousand men and girls, mostly girls, - as many girls as would fill Vassar College a hundred times and more, or double the population of Nevada. Put these men and girls in line, march them ten abreast, and six hours would pass before the last company would arrive at the reviewing stand. In single file this throng of Telephonians would make a living wall from New York to New Haven.

Such is the extraordinary city of which Alexander Graham Bell was the only resident in 1875. It has been built up without the backing of any great bank or multi-millionaire. There have been no Vanderbilts in it, no Astors, Rockefellers, Rothschilds, Harrimans. There are even now only four men who own as many as ten thousand shares of the stock of the central company. This Bell System stands as the life-work of unprivileged men, who are for the most part still alive and busy. With very few and trivial exceptions, every part of it was made in the United States. No other industrial organism of equal size owes foreign countries so little. Alike in its origin, its development, and its highest point of efficiency and expansion, the telephone is as essentially American as the Declaration of Independence or the monument on Bunker Hill.

CHAPTER VI

NOTABLE USERS OF THE TELEPHONE

What we might call the telephonization of city life, for lack of a simpler word, has remarkably altered our manner of living from what it was in the days of Abraham Lincoln. It has enabled us to be more social and cooperative. It has literally abolished the isolation of separate families, and has made us members of one great family. It has become so truly an organ of the social body that by telephone we now enter into contracts, give evidence, try lawsuits, make speeches, propose marriage, confer degrees, appeal to voters, and do almost everything else that is a matter of speech.

In stores and hotels this wire traffic has grown to an almost bewildering extent, as these are the places where many interests meet. The hundred largest hotels in New York City have twenty-one thousand telephones - nearly as many as the continent of Africa and more than the kingdom of Spain. In an average year they send six million messages. The Waldorf-Astoria alone tops all residential buildings with eleven hundred and twenty telephones and five hundred thousand calls a year; while merely the Christmas Eve orders that flash into Marshall Field's store, or John Wanamaker's, have risen as high as the three thousand mark.

Whether the telephone does most to concentrate population, or to scatter it, is a question that has not yet been examined. It is certainly true that it has made the skyscraper possible, and thus helped to create an absolutely new type of city, such as was never imagined even in the fairy tales of ancient nations. The skyscraper is ten years younger than the telephone. It is now generally seen to be the ideal building for business offices. It is one of the few types of architecture that may fairly be called American. And its efficiency is largely, if not mainly, due to the fact that its inhabitants may run errands by telephone as well as by elevator.

There seems to be no sort of activity which is not being made more convenient by the telephone. It is used to call the duck-shooters in Western Canada when a flock of birds has arrived; and to direct the movements of the Dragon in Wagner's grand opera "Siegfried." At the last Yale-Harvard football game, it conveyed almost instantaneous news to fifty thousand people in various parts of New England. At the Vanderbilt Cup Race its wires girdled the track and reported every gain or mishap of the racing autos. And at such expensive pageants as that of the Quebec Tercentenary in 1908, where four thousand actors came and went upon a ten-acre stage, every order was given by telephone.

Public officials, even in the United States, have been slow to change from the old-fashioned and more dignified use of written documents and uniformed messengers; but in the last ten years there has been a sweeping revolution in this respect. Government by telephone! This is a new idea that has already arrived in the more efficient departments of the Federal service. And as for the present Congress, that body has gone so far as to plan for a special system of its own, in both Houses, so that all official announcements may be heard by wire.

Garfield was the first among American Presidents to possess a telephone. An exhibition instrument was placed in his house, without cost, in 1878, while he was still a member of Congress. Neither Cleveland nor Harrison, for temperamental reasons, used the magic wire very often. Under their regime, there was one lonely idle telephone in the White House, used by the servants several times a week. But with McKinley came a new order of things. To him a telephone was more than a necessity. It was a pastime, an exhilarating sport. He was the one President who really revelled in the comforts of telephony. In 1895 he sat in his Canton home and heard the cheers of the Chicago Convention. Later he sat there and ran the first presidential telephone campaign; talked to his managers in thirty-eight States. Thus he came to regard the telephone with a higher degree of appreciation than any of his predecessors had done, and eulogized it on many public occasions. "It is bringing us all closer together," was his favorite phrase.

To Roosevelt the telephone was mainly for emergencies. He used it to the full during the Chicago Convention of 1907 and the Peace Conference at Portsmouth. But with Taft the telephone became again the common avenue of conversation. He has introduced at least one new telephonic custom a long-distance talk with his family every evening, when he is away from home. Instead of the solitary telephone of Cleveland-Harrison days, the White House has now a branch exchange of its own - Main 6 - with a sheaf of wires that branch out into every room as well as to the nearest central.

Next to public officials, bankers were perhaps the last to accept the facilities of the telephone. They were slow to abandon the fallacy that no business can be done without a written record. James Stillman, of New York, was first among bankers to foresee the telephone era. As early as 1875, while Bell was teaching his infant telephone to talk, Stillman risked two thousand dollars in a scheme to establish a crude dial system of wire communication, which later grew into New York's first telephone exchange. At the present time, the banker who works closest to his telephone is probably George W. Perkins, of the J. P. Morgan group of bankers. "He is the only man," says Morgan, "who can raise twenty millions in twenty minutes." The Perkins plan of rapid transit telephony is to prepare a list of names, from ten to thirty, and to flash from one to another as fast as the operator can ring them up. Recently one of the other members of the Morgan bank proposed to enlarge its telephone equipment. "What will we gain by more wires?" asked the operator. "If we were to put in a six-hundred pair cable, Mr. Perkins would keep it busy."

The most brilliant feat of the telephone in the financial world was done during the panic of 1907. At the height of the storm, on a Saturday evening, the New York bankers met in an almost desperate conference. They decided, as an emergency measure of self-protection, not to ship cash to Western banks. At midnight they telephoned this decision to the bankers of Chicago and St. Louis. These men, in turn, conferred by telephone, and on Sunday afternoon called up the bankers of neighboring States. And so the news went from 'phone to 'phone, until by Monday morning all bankers and chief depositors were aware of the situation, and prepared for the team-play that prevented any general disaster.

As for stockbrokers of the Wall Street species, they transact practically all their business by telephone. In their stock exchange stand six hundred and forty one booths, each one the terminus of a private wire. A firm of brokers will count it an ordinary year's talking to send fifty thousand messages; and there is one firm which last year sent twice as many. Of all brokers, the one who finally accomplished most by telephony was unquestionably E. H. Harriman. In the mansion that he built at Arden, there were a hundred telephones, sixty of them linked to the long-distance lines. What the brush is to the artist, what the chisel is to the sculptor, the telephone was to Harriman. He built his fortune with it. It was in his library, his bathroom, his private car, his camp in the Oregon wilderness. No transaction was too large or too involved to be settled over its wires. He saved the credit of the Erie by telephone - lent it five million dollars as he lay at home on a sickbed. "He is a slave to the telephone," wrote a magazine writer. "Nonsense," replied Harriman, "it is a slave to me."

The telephone arrived in time to prevent big corporations from being unwieldy and aristocratic. The foreman of a Pittsburg coal company may now stand in his subterranean office and talk to the president of the Steel Trust, who sits on the twenty-first floor of a New York skyscraper. The long-distance talks, especially, have grown to be indispensable to the corporations whose plants are scattered and geographically misplaced - to the mills of New England, for instance, that use the cotton of the South and sell so much of their product to the Middle West. To the companies that sell perishable commodities, an instantaneous conversation with a buyer in a distant city has often saved a carload or a cargo. Such caterers as the meat-packers, who were among the first to realize what Bell had made possible, have greatly accelerated the wheels of their business by inter-city conversations. For ten years or longer the Cudahys have talked every business morning between Omaha and Boston, via fifteen hundred and seventy miles of wire.

In the refining of oil, the Standard Oil Company alone, at its New York office, sends two hundred and thirty thousand messages a year. In the making of steel, a chemical analysis is made of each caldron of molten pig-iron, when it starts on its way to be refined, and this analysis is sent by telephone to the steelmaker, so that he will know exactly how each potful is to be handled. In the floating of logs down rivers, instead of having relays of shouters to prevent the logs from jamming, there is now a wire along the bank, with a telephone linked on at every point of danger. In the rearing of skyscrapers, it is now usual to have a temporary wire strung vertically, so that the architect may stand on the ground and confer with a foreman who sits astride of a naked girder three hundred feet up in the air. And in the electric light business, the current is distributed wholly by telephoned orders. To give New York the seven million electric lights that have abolished night in that city requires twelve private exchanges and five hundred and twelve telephones. All the power that creates this artificial daylight is generated at a single station, and let flow to twenty-five storage centres. Minute by minute, its flow is guided by an expert, who sits at a telephone exchange as though he were a pilot at the wheel of an ocean liner.

The first steamship line to take notice of the telephone was the Clyde, which had a wire from dock to office in 1877; and the first railway was the Pennsylvania, which two years later was persuaded by Professor Bell himself to give it a trial in Altoona. Since then, this railroad has become the chief beneficiary of the art of telephony. It has one hundred and seventy-five exchanges, four hundred operators, thirteen thousand telephones, and twenty thousand miles of wire - a more ample system than the city of New York had in 1896.

To-day the telephone goes to sea in the passenger steamer and the warship. Its wires are waiting at the dock and the depot, so that a tourist may sit in his stateroom and talk with a friend in some distant office. It is one of the most incredible miracles of telephony that a passenger at New York, who is about to start for Chicago on a fast express, may telephone to Chicago from the drawing-room of a Pullman. He himself, on the swiftest of all trains, will not arrive in Chicago for eighteen hours; but the flying words can make the journey, and RETURN, while his train is waiting for the signal to start.

In the operation of trains, the railroads have waited thirty years before they dared to trust the telephone, just as they waited fifteen years before they dared to trust the telegraph. In 1883 a few railways used the telephone in a small way, but in 1907, when a law was passed that made telegraphers highly expensive, there was a general swing to the telephone. Several dozen roads have now put it in use, some employing it as an associate of the Morse method and others as a complete substitute. It has already been found to be the quickest way of despatching trains. It will do in five minutes what the telegraph did in ten. And it has enabled railroads to hire more suitable men for the smaller offices.

In news-gathering, too, much more than in railroading, the day of the telephone has arrived. The Boston Globe was the first paper to receive news by telephone. Later came The Washington Star, which had a wire strung to the Capitol, and thereby gained an hour over its competitors. To-day the evening papers receive most of their news over the wire a la Bell instead of a la Morse. This has resulted in a specialization of reporters - one man runs for the news and another man writes it. Some of the runners never come to the office. They receive their assignments by telephone, and their salaries by mail. There are even a few who are allowed to telephone their news directly to a swift linotype operator, who clicks it into type on his machine, without the scratch of a pencil. This, of course, is the ideal method of news-gathering, which is rarely possible.

A paper of the first class, such as The New York World, has now an outfit of twenty trunk lines and eighty telephones. Its outgoing calls are two hundred thousand a year and its incoming calls three hundred thousand, which means that for every morning, evening, or Sunday edition, there has been an average of seven hundred and fifty messages. The ordinary newspaper in a small town cannot afford such a service, but recently the United Press has originated a cooperative method. It telephones the news over one wire to ten or twelve newspapers at one time. In ten minutes a thousand words can in this way be flung out to a dozen towns, as quickly as by telegraph and much cheaper.

But it is in a dangerous crisis, when safety seems to hang upon a second, that the telephone is at its best. It is the instrument of emergencies, a sort of ubiquitous watchman. When the girl operator in the exchange hears a cry for help - "Quick! The hospital!" "The fire department!" "The police!" she seldom waits to hear the number. She knows it. She is trained to save half-seconds. And it is at such moments, if ever, that the users of a telephone can appreciate its insurance value. No doubt, if a King Richard III were worsted on a modern battlefield, his instinctive cry would be, "My Kingdom for a telephone!"

When instant action is needed in the city of New York, a General Alarm can in five minutes be sent by the police wires over its whole vast area of three hundred square miles. When, recently, a gas main broke in Brooklyn, sixty girls were at once called to the centrals in that part of the city to warn the ten thousand families who had been placed in danger. When the ill-fated General Slocum caught fire, a mechanic in a factory on the water-front saw the blaze, and had the presence of mind to telephone the newspapers, the hospitals, and the police. When a small child is lost, or a convict has escaped from prison, or the forest is on fire, or some menace from the weather is at hand, the telephone bells clang out the news, just as the nerves jangle the bells of pain when the body is in danger. In one tragic case, the operator in Folsom, New Mexico, refused to quit her post until she had warned her people of a flood that had broken loose in the hills above the village. Because of her courage, nearly all were saved, though she herself was drowned at the switchboard. Her name - Mrs. S. J. Rooke - deserves to be remembered.

If a disaster cannot be prevented, it is the telephone,

usually, that brings first aid to the injured. After the destruction of San Francisco, Governor Guild, of Massachusetts, sent an appeal for the stricken city to the three hundred and fifty-four mayors of his State; and by the courtesy of the Bell Company, which carried the messages free, they were delivered to the last and furthermost mayors in less than five hours. After the destruction of Messina, an order for enough lumber to build ten thousand new houses was cabled to New York and telephoned to Western lumbermen. So quickly was this order filled that on the twelfth day after the arrival of the cablegram, the ships were on their way to Messina with the lumber. After the Kansas City flood of 1903, when the drenched city was without railways or street-cars or electric lights, it was the telephone that held the city together and brought help to the danger-spots. And after the Baltimore fire, the telephone exchange was the last force to quit and the first to recover. Its girls sat on their stools at the switchboard until the window-panes were broken by the heat. Then they pulled the covers over the board and walked out. Two hours later the building was in ashes. Three hours later another building was rented on the unburned rim of the city, and the wire chiefs were at work. In one day there was a system of wires for the use of the city officials. In two days these were linked to long-distance wires; and in eleven days a two-thousand-line switchboard was in full working trim. This feat still stands as the record in rebuilding.

In the supreme emergency of war, the telephone is as indispensable, very nearly, as the cannon. This, at least, is the belief of the Japanese, who handled their armies by telephone when they drove back the Russians. Each body of Japanese troops moved forward like a silkworm, leaving behind it a glistening strand of red copper wire. At the decisive battle of Mukden, the silk-worm army, with a million legs, crept against the Russian hosts in a vast crescent, a hundred miles from end to end. By means of this glistening red wire, the various batteries and regiments were organized into fifteen divisions. Each group of three divisions was wired to a general, and the five generals were wired to the great Oyama himself, who sat ten miles back of the firing-line and sent his orders. Whenever a regiment lunged forward, one of the soldiers carried a telephone set. If they held their position, two other soldiers ran forward with a spool of wire. In this way and under fire of the Russian cannon, one hundred and fifty miles of wire were strung across the battlefield. As the Japanese said, it was this "flying telephone" that enabled Oyama to manipulate his forces as handily as though he were playing a game of chess. It was in this war, too, that the Mikado's soldiers strung the costliest of all telephone lines, at 203 Metre Hill. When the wire had been basted up this hill to the summit, the fortress of Port Arthur lay at their mercy. But the climb had cost them twenty-four thousand lives.

Of the seven million telephones in the United States, about two million are now in farmhouses. Every fourth American farmer is in telephone touch with his neighbors and the market. Iowa leads, among the farming States. In Iowa, not to have a telephone is to belong to what a Londoner would call the "submerged tenth" of the population. Second in line comes Illinois, with Kansas, Nebraska, and Indiana following closely behind; and at the foot of the list, in the matter of farm telephones, are Connecticut and Louisiana.

The first farmer who discovered the value of the

telephone was the market gardener. Next came the bonanza farmer of the Red River Valley - such a man, for instance, as Oliver Dalrymple, of North Dakota, who found that by the aid of the telephone he could plant and harvest thirty thousand acres of wheat in a single season. Then, not more than half a dozen years ago, there arose a veritable Telephone Crusade among the farmers of the Middle West. Cheap telephones, yet fairly good, had by this time been made possible by the improvements of the Bell engineers; and stories of what could be done by telephone became the favorite gossip of the day. One farmer had kept his barn from being burned down by telephoning for his neighbors; another had cleared five hundred dollars extra profit on the sale of his cattle, by telephoning to the best market; a third had rescued a flock of sheep by sending quick news of an approaching blizzard; a fourth had saved his son's life by getting an instantaneous message to the doctor; and so on.

How the telephone saved a three million dollar fruit crop in Colorado, in 1909, is the story that is oftenest told in the West. Until that year, the frosts in the Spring nipped the buds. No farmer could be sure of his harvest. But in 1909, the fruit-growers bought smudge-pots - three hundred thousand or more. These were placed in the orchards, ready to be lit at a moment's notice. Next, an alliance was made with the United States Weather Bureau so that whenever the Frost King came down from the north, a warning could be telephoned to the farmers. Just when Colorado was pink with apple blossoms, the first warning came. "Get ready to light up your smudge-pots in half an hour." Then the farmers telephoned to the nearest towns: "Frost is coming; come and help us in the orchards." Hundreds of men rushed out into the country on horseback and in wagons. In half an hour the last warning came: "Light up; the thermometer registers twenty-nine." The smudge-pot artillery was set ablaze, and kept blazing until the news came that the icy forces had retreated. And in this way every Colorado farmer who had a telephone saved his fruit.

In some farming States, the enthusiasm for the telephone is running so high that mass meetings are held, with lavish oratory on the general theme of "Good Roads and Telephones." And as a result of this Telephone Crusade, there are now nearly twenty thousand groups of farmers, each one with a mutual telephone system, and one-half of them with sufficient enterprise to link their little webs of wires to the vast Bell system, so that at least a million farmers have been brought as close to the great cities as they are to their own barns.

What telephones have done to bring in the present era of big crops, is an interesting story in itself. To compress it into a sentence, we might say that the telephone has completed the labor-saving movement which started with the McCormick reaper in 1831. It has lifted the farmer above the wastefulness of being his own errand-boy. The average length of haul from barn to market in the United States is nine and a half miles, so that every trip saved means an extra day's work for a man and team. Instead of travelling back and forth, often to no purpose, the farmer may now stay at home and attend to his stock and his crops.

As yet, few farmers have learned to appreciate the value of quality in telephone service, as they have in other lines. The same man who will pay six prices for the best seed-corn, and who will allow nothing but high-grade cattle in his barn, will at the same time be content with the shabbiest and flimsiest telephone service, without offering any other excuse than that it is cheap. But this is a transient phase of farm telephony. The cost of an efficient farm system is now so little - not more than two dollars a month, that the present trashy lines are certain sooner or later to go to the junk-heap with the sickle and the flail and all the other cheap and unprofitable things.

CHAPTER VII

THE TELEPHONE AND NATIONAL EFFICIENCY

The larger significance of the telephone is that it completes the work of eliminating the hermit and gypsy elements of civilization. In an almost ideal way, it has made intercommunication possible without travel. It has enabled a man to settle permanently in one place, and yet keep in personal touch with his fellows.

Until the last few centuries, much of the world was probably what Morocco is to-day - a region without wheeled vehicles or even roads of any sort. There is a mythical story of a wonderful speaking-trumpet possessed by Alexander the Great, by which he could call a soldier who was ten miles distant; but there was probably no substitute for the human voice except flags and beacon-fires, or any faster method of travel than the gait of a horse or a camel across ungraded plains. The first sensation of rapid transit doubtless came with the sailing vessel; but it was the play-toy of the winds, and unreliable. When Columbus dared to set out on his famous voyage, he was five weeks in crossing from Spain to the West Indies, his best day's record two hundred miles. The swift steamship travel of to-day did not begin until 1838, when the Great Western raced over the Atlantic in fifteen days.

As for organized systems of intercommunication, they were unknown even under the rule of a Pericles or a Caesar. There was no post office in Great Britain until 1656 - a generation after America had begun to be colonized. There was no English mail-coach until 1784; and when Benjamin Franklin was Postmaster General at Philadelphia, an answer by mail from Boston, when all went well, required not less than three weeks. There was not even a hard-surface road in the thirteen United States until 1794; nor even a postage stamp until 1847, the year in which Alexander Graham Bell was born. In this same year Henry Clay delivered his memorable speech on the Mexican War, at Lexington, Kentucky, and it was telegraphed to The New York Herald at a cost of five hundred dollars, thus breaking all previous records for news-gathering enterprise. Eleven years later the first cable established an instantaneous sign-language between Americans and Europeans; and in 1876 there came the perfect distance-talking of the telephone.

No invention has been more timely than the telephone. It arrived at the exact period when it was needed for the organization of great cities and the unification of nations. The new ideas and energies of science, commerce, and cooperation were beginning to win victories in all parts of the earth. The first railroad had just arrived in China; the first parliament in Japan; the first constitution in Spain. Stanley was moving like a tiny point of light through the heart of the Dark Continent. The Universal Postal Union had been organized in a little hall in Berne. The Red Cross movement was twelve years old. An International Congress of Hygiene was being held at Brussels, and an International Congress of Medicine at Philadelphia. De Lesseps had finished the Suez Canal and was examining Panama. Italy and Germany had recently been built into nations; France had finally swept aside the Empire and the Commune and established the Republic. And what with the new agencies of railroads, steamships, cheap newspapers, cables, and telegraphs, the civilized races of mankind had begun to be knit together into a practical consolidation.

To the United States, especially, the telephone came as a friend in need. After a hundred years of growth, the Republic was still a loose confederation of separate States, rather than one great united nation. It had recently fallen apart for four years, with a wide gulf of blood between; and with two flags, two Presidents, and two armies. In 1876 it was hesitating halfway between doubt and confidence, between the old political issues of North and South, and the new industrial issues of foreign trade and the development of material resources. The West was being thrown open. The Indians and buffaloes were being driven back. There was a line of railway from ocean to ocean. The population was gaining at the rate of a million a year. Colorado had just been baptized as a new State. And it was still an unsolved problem whether or not the United States could be kept united, whether or not it could be built into an organic nation without losing the spirit of self-help and democracy.

It is not easy for us to realize to-day how young and primitive was the United States of 1876. Yet the fact is that we have twice the population that we had when the telephone was invented. We have twice the wheat crop and twice as much money in circulation. We have three times the railways, banks, libraries, newspapers, exports, farm values, and national wealth. We have ten million farmers who make four times as much money as seven million farmers made in 1876. We spend four times as much on our public schools, and we put four times as much in the savings bank. We have five times as many students in the colleges. And we have so revolutionized our methods of production that we now produce seven times as much coal, fourteen times as much oil and pig-iron, twenty-two times as much copper, and forty-three times as much steel.

There were no skyscrapers in 1876, no trolleys, no electric lights, no gasoline engines, no self-binders, no bicycles, no automobiles. There was no Oklahoma, and the combined population of Montana, Wyoming, Idaho, and Arizona was about equal to that of Des Moines. It was in this year that General Custer was killed by the Sioux; that the flimsy iron railway bridge fell at Ashtabula; that the "Molly Maguires" terrorized Pennsylvania; that the first wire of the Brooklyn Bridge was strung; and that Boss Tweed and Hell Gate were both put out of the way in New York.

The Great Elm, under which the Revolutionary patriots had met, was still standing on Boston Common. Daniel Drew, the New York financier, who was born before the American Constitution was adopted, was still alive; so were Commodore Vanderbilt, Joseph Henry, A. T. Stewart, Thurlow Weed, Peter Cooper, Cyrus McCormick, Lucretia Mott, Bryant, Longfellow, and Emerson. Most old people could remember the running of the first railway train; people of middle age could remember the sending of the first telegraph message; and the children in the high schools remembered the laying of the first Atlantic Cable.

The grandfathers of 1876 were fond of telling how Webster opposed taking Texas and Oregon into the

Union; how George Washington advised against including the Mississippi River; and how Monroe warned Congress that a country that reached from the Atlantic to the Middle West was "too extensive to be governed but by a despotic monarchy." They told how Abraham Lincoln, when he was postmaster of New Salem, used to carry the letters in his coon-skin cap and deliver them at sight; how in 1822 the mails were carried on horseback and not in stages, so as to have the quickest possible service; and how the news of Madison's election was three weeks in reaching the people of Kentucky. When the telegraph was mentioned, they told how in Revolutionary days the patriots used a system of signalling called "Washington's Telegraph," consisting of a pole, a flag, a basket, and a barrel.

So, the young Republic was still within hearing distance of its childhood, in 1876. Both in sentiment and in methods of work it was living close to the log-cabin period. Many of the old slow ways survived, the ways that were fast enough in the days of the stage-coach and the tinder-box. There were seventy-seven thousand miles of railway, but poorly built and in short lengths. There were manufacturing industries that employed two million, four hundred thousand people, but every trade was broken up into a chaos of small competitive units, each at war with all the others. There were energy and enterprise in the highest degree, but not efficiency or organization. Little as we knew it, in 1876 we were mainly gathering together the plans and the raw materials for the building up of the modern business world, with its quick, tense life and its national structure of immense coordinated industries.

In 1876 the age of specialization and community of

interest was in its dawn. The cobbler had given place to the elaborate factory, in which seventy men cooperated to make one shoe. The merchant who had hitherto lived over his store now ventured to have a home in the suburbs. No man was any longer a self-sufficient Robinson Crusoe. He was a fraction, a single part of a social mechanism, who must necessarily keep in the closest touch with many others.

A new interdependent form of civilization was about to be developed, and the telephone arrived in the nick of time to make this new civilization workable and convenient. It was the unfolding of a new organ. Just as the eye had become the telescope, and the hand had become machinery, and the feet had become railways, so the voice became the telephone. It was a new ideal method of communication that had been made indispensable by new conditions. The prophecy of Carlyle had come true, when he said that "men cannot now be bound to men by brass collars; you will have to bind them by other far nobler and cunninger methods."

Railways and steamships had begun this work of binding man to man by "nobler and cunninger methods." The telegraph and cable had gone still farther and put all civilized people within sight of each other, so that they could communicate by a sort of deaf and dumb alphabet. And then came the telephone, giving direct instantaneous communication and putting the people of each nation within hearing distance of each other. It was the completion of a long series of inventions. It was the keystone of the arch. It was the one last improvement that enabled interdependent nations to handle themselves and to hold together.

To make railways and steamboats carry letters was

much, in the evolution of the means of communication. To make the electric wire carry signals was more, because of the instantaneous transmission of important news. But to make the electric wire carry speech was MOST, because it put all fellow-citizens face to face, and made both message and answer instantaneous. The invention of the telephone taught the Genie of Electricity to do better than to carry messages in the sign language of the dumb. It taught him to speak. As Emerson has finely said:

"We had letters to send. Couriers could not go fast enough, nor far enough; broke their wagons, foundered their horses; bad roads in Spring, snowdrifts in Winter, heat in Summer - could not get their horses out of a walk. But we found that the air and the earth were full of electricity, and always going our way, just the way we wanted to send. WOULD HE TAKE A MESSAGE, Just as lief as not; had nothing else to do; would carry it in no time."

As to the exact value of the telephone to the United States in dollars and cents, no one can tell. One statistician has given us a total of three million dollars a day as the amount saved by using telephones. This sum may be far too high, or too low. It can be no more than a guess. The only adequate way to arrive at the value of the telephone is to consider the nation as a whole, to take it all in all as a going concern, and to note that such a nation would be absolutely impossible without its telephone service. Some sort of a slower and lower grade republic we might have, with small industrial units, long hours of labor, lower wages, and clumsier ways. The money loss would be enormous, but more serious still would be the loss in the QUALITY OF THE NATIONAL LIFE. Inevitably, an untelephoned nation is less social, less unified, less progressive, and less efficient. It belongs to an inferior species.

How to make a civilization that is organized and quick, instead of a barbarism that was chaotic and slow - that is the universal human problem, not wholly solved today. And how to develop a science of intercommunication, which commenced when the wild animals began to travel in herds and to protect themselves from their enemies by a language of danger-signals, and to democratize this science until the entire nation becomes self-conscious and able to act as one living being - that is the part of this universal problem which finally necessitated the invention of the telephone.

With the use of the telephone has come a new habit of mind. The slow and sluggish mood has been sloughed off. The old to-morrow habit has been superseded by "Do It To-day"; and life has become more tense, alert, vivid. The brain has been relieved of the suspense of waiting for an answer, which is a psychological gain of great importance. It receives its reply at once and is set free to consider other matters. There is less burden upon the memory and the WHOLE MIND can be given to each new proposition.

A new instinct of speed has been developed, much more fully in the United States than elsewhere. "No American goes slow," said Ian Maclaren, "if he has the chance of going fast; he does not stop to talk if he can talk walking; and he does not walk if he can ride." He is as pleased as a child with a new toy when some speed record is broken, when a pair of shoes is made in eleven minutes, when a man lays twelve hundred bricks in an hour, or when a ship crosses the Atlantic in four and a half days. Even seconds are now counted and split up into fractions. The average time, for instance, taken to reply to a telephone call by a New York operator, is now three and two-fifth seconds; and even this tiny atom of time is being strenuously worn down.

As a witty Frenchman has said, one of our most lively regrets is that while we are at the telephone we cannot do business with our feet. We regard it as a victory over the hostility of nature when we do an hour's work in a minute or a minute's work in a second. Instead of saying, as the Spanish do, "Life is too short; what can one person do?" an American is more apt to say, "Life is too short; therefore I must do to-day's work to-day." To pack a lifetime with energy - that is the American plan, and so to economize that energy as to get the largest results. To get a question asked and answered in five minutes by means of an electric wire, instead of in two hours by the slow trudging of a messenger boy - that is the method that best suits our passion for instantaneous service.

It is one of the few social laws of which we are fairly sure, that a nation organizes in proportion to its velocity. We know that a four-mile-an-hour nation must remain a huge inert mass of peasants and villagers; or if, after centuries of slow toil, it should pile up a great city, the city will sooner or later fall to pieces of its own weight. In such a way Babylon rose and fell, and Nineveh, and Thebes, and Carthage, and Rome. Mere bulk, unorganized, becomes its own destroyer. It dies of clogging and congestion. But when Stephenson's Rocket ran twenty-nine miles an hour, and Morse's telegraph clicked its signals from Washington to Baltimore, and Bell's telephone flashed the vibrations of speech between Boston and Salem, a new era began. In came the era of speed and the finely organized nations. In came cities of unprecedented bulk, but held together so closely by a web-work of steel rails and copper wires that they have become more alert and cooperative than any tiny hamlet of mud huts on the banks of the Congo.

That the telephone is now doing most of all, in this binding together of all manner of men, is perhaps not too much to claim, when we remember that there are now in the United States seventy thousand holders of Bell telephone stock and ten million users of telephone service. There are two hundred and sixty-four wires crossing the Mississippi, in the Bell system; and five hundred and forty-four crossing Mason and Dixon's Line. It is the telephone which does most to link together cottage and skyscraper and mansion and factory and farm. It is not limited to experts or college graduates. It reaches the man with a nickel as well as the man with a million. It speaks all languages and serves all trades. It helps to prevent sectionalism and race feuds. It gives a common meeting place to capitalists and wage-workers. It is so essentially the instrument of all the people, in fact, that we might almost point to it as a national emblem, as the trade-mark of democracy and the American spirit.

In a country like ours, where there are eighty nationalities in the public schools, the telephone has a peculiar value as a part of the national digestive apparatus. It prevents the growth of dialects and helps on the process of assimilation. Such is the push of American life, that the humble immigrants from Southern Europe, before they have been here half a dozen years, have acquired the telephone habit and have linked on their small shops to the great wire network of intercommunication. In the one community of Brownsville, for example, settled several years ago by an overflow of Russian Jews from the East Side of New York, there are now as many telephones as in the kingdom of Greece. And in the swarming East Side itself, there is a single exchange in Orchard Street which has more wires than there are in all the exchanges of Egypt.

There can be few higher ideals of practical democracy than that which comes to us from the telephone engineer. His purpose is much more comprehensive than the supplying of telephones to those who want them. It is rather to make the telephone as universal as the water faucet, to bring within speaking distance every economic unit, to connect to the social organism every person who may at any time be needed. Just as the click of the reaper means bread, and the purr of the sewing-machine means clothes, and the roar of the Bessemer converter means steel, and the rattle of the press means education, so the ring of the telephone bell has come to mean unity and organization.

Already, by cable, telegraph, and telephone, no two towns in the civilized world are more than one hour apart. We have even girdled the earth with a cablegram in twelve minutes. We have made it possible for any man in New York City to enter into conversation with any other New Yorker in twenty-one seconds. We have not been satisfied with establishing such a system of transportation that we can start any day for anywhere from anywhere else; neither have we been satisfied with establishing such a system of communication that news and gossip are the common property of all nations. We have gone farther. We have established in every large region of population a system of voice-nerves that puts every man at every other man's ear, and which so magically eliminates the factor of distance that the United States becomes three thousand miles of neighbors, side by side.

This effort to conquer Time and Space is above all else the instinct of material progress. To shrivel up the miles and to stretch out the minutes - this has been one of the master passions of the human race. And thus the larger truth about the telephone is that it is vastly more than a mere convenience. It is not to be classed with safety razors and piano players and fountain pens. It is nothing less than the high-speed tool of civilization, gearing up the whole mechanism to more effective social service. It is the symbol of national efficiency and coperation.

All this the telephone is doing, at a total cost to the nation of probably $200,000,000 a year - no more than American farmers earn in ten days. We pay the same price for it as we do for the potatoes, or for one-third of the hay crop, or for one-eighth of the corn. Out of every nickel spent for electrical service, one cent goes to the telephone. We could settle our telephone bill, and have several millions left over, if we cut off every fourth glass of liquor and smoke of tobacco. Whoever rents a typewriting machine, or uses a street car twice a day, or has his shoes polished once a day, may for the same expense have a very good telephone service. Merely to shovel away the snow of a single storm in 1910 cost the city government of New York as much as it will pay for five or six years of telephoning.

This almost incredible cheapness of telephony is still far from being generally perceived, mainly for psychological reasons. A telephone is not impressive. It has no bulk. It is not like the Singer Building or the Lusitania. Its wires and switchboards and batteries are scattered and hidden, and few have sufficient imagination to picture them in all their complexity. If only it were possible to assemble the hundred or more telephone buildings of New York in one vast plaza, and if the two thousand clerks and three thousand maintenance men and six thousand girl operators were to march to work each morning with bands and banners, then, perhaps, there might be the necessary quality of impressiveness by which any large idea must always be imparted to the public mind.

For lack of a seven and one-half cent coin, there is now five-cent telephony even in the largest American cities. For five cents whoever wishes has an entire wire-system at his service, a system that is kept waiting by day and night, so that it will be ready the instant he needs it. This system may have cost from twenty to fifty millions, yet it may be hired for one-eighth the cost of renting an automobile. Even in long-distance telephony, the expense of a message dwindles when it is compared with the price of a return railway ticket. A talk from New York to Philadelphia, for instance, costs seventy-five cents, while the railway fare would be four dollars. From New York to Chicago a talk costs five dollars as against seventy dollars by rail. As Harriman once said, "I can't get from my home to the depot for the price of a talk to Omaha."

To say what the net profits have been, to the entire body of people who have invested money in the telephone, will always be more or less of a guess. The general belief that immense fortunes were made by the lucky holders of Bell stock, is an exaggeration that has been kept alive by the promoters of wildcat companies. No such fortunes were made. "I do not believe," says Theodore Vail, "that any one man ever made a clear million out of the telephone." There are not apt to be any get-rich-quick for-tunes made in corporations that issue no watered stock and do not capitalize their franchises. On the contrary, up to 1897, the holders of stock in the Bell Companies had paid in four million, seven hundred thousand dollars more than the par value; and in the recent consolidation of Eastern companies, under the presidency of Union N. Bethell, the new stock was actually eight millions less than the stock that was retired.

Few telephone companies paid any profits at first. They had undervalued the cost of building and maintenance. Denver expected the cost to be two thousand, five hundred dollars and spent sixty thousand dollars. Buffalo expected to pay three thousand dollars and had to pay one hundred and fifty thousand dollars. Also, they made the unwelcome discovery that an exchange of two hundred costs more than twice as much as an exchange of one hundred, because of the greater amount of traffic. Usually a dollar that is paid to a telephone company is divided as follows:

Rent 4c
Taxes 4c
Interest 6c
Surplus 8c
Maintenance 16c
Dividends 18c
Labor 44c
- -
$1.00

Most of the rate troubles (and their name has been legion) have arisen because the telephone business was not understood. In fact, until recently, it did not understand itself. It persisted in holding to a local and individualistic view of its business. It was slow to put telephones in unprofitable places. It expected every instrument to pay its way. In many States, both the telephone men and the public overlooked the most vital fact in the case, which is that the members of a telephone system are above all else INTERDEPENDENT.

One telephone by itself has no value. It is as useless as a reed cut out of an organ or a finger that is severed from a hand. It is not even ornamental or adaptable to any other purpose. It is not at all like a piano or a talking-machine, which has a separate existence. It is useful only in proportion to the number of other telephones it reaches. AND EVERY TELEPHONE ANYWHERE ADDS VALUE TO EVERY OTHER TELEPHONE ON THE SAME SYSTEM OF WIRES. That, in a sentence, is the keynote of equitable rates.

Many a telephone, for the general good, must be put where it does not earn its own living. At any time some sudden emergency may arise that will make it for the moment priceless. Especially since the advent of the automobile, there is no nook or corner from which it may not be supremely necessary, now and then, to send a message. This principle was acted upon recently in a most practical way by the Pennsylvania Railroad, which at its own expense installed five hundred and twenty-five telephones in the homes of its workmen in Altoona. In the same way, it is clearly the social duty of the telephone company to widen out its system until every point is covered, and then to distribute its gross charges as fairly as it can. The whole must carry the whole - that is the philosophy of rates which must finally be recognized by legislatures and telephone companies alike. It can never, of course, be reduced to a system or formula. I* will always be a matter of opinion and compromise, requiring much skill and much patience. But there will seldom be any serious trouble when once its basic principles are understood.

Like all time-saving inventions, like the railroad, the reaper, and the Bessemer converter, the telephone, in the last analysis, COSTS NOTHING; IT IS THE LACK OF IT THAT COSTS. THE NATION THAT MOST IS THE NATION WITHOUT IT.

CHAPTER VIII

THE TELEPHONE IN FOREIGN COUNTRIES

The telephone was nearly a year old before Europe was aware of its existence. It received no public notice of any kind whatever until March 3, 1877, when the London Athenaeum mentioned it in a few careful sentences. It was not welcomed, except by those who wished an evening's entertainment. And to the entire commercial world it was for four or five years a sort of scientific Billiken, that never could be of any service to serious people.

One after another, several American enthusiasts rushed posthaste to Europe, with dreams of eager nations clamoring for telephone systems, and one after another they failed. Frederick A. Gower was the first of these. He was an adventurous chevalier of business who gave up an agent's contract in return for a right to become a roving propagandist. Later he met a prima donna, fell in love with and married her, forsook telephony for ballooning, and lost his life in attempting to fly across the English Channel.

Next went William H. Reynolds, of Providence, who had bought five-eights of the British patent for five thousand dollars, and half the right to Russia, Spain, Portugal, and Italy for two thousand, five hundred dollars. How he was received may be seen from a letter of his which has been preserved. "I have been working in London for four months," he writes; "I have been to the Bank of England and elsewhere; and I have not found one man who will put one shilling into the telephone."

Bell himself hurried to England and Scotland on his wedding tour in 1878, with great expectations of having his invention appreciated in his native land. But from a business point of view, his mission was a total failure. He received dinners a-plenty, but no contracts; and came back to the United States an impoverished and disheartened man. Then the optimistic Gardiner G. Hubbard, Bell's father-in-law, threw himself against the European inertia and organized the International and Oriental Telephone Companies, which came to nothing of any importance. In the same year even Enos M. Barton, the sagacious founder of the Western Electric, went to France and England to establish an export trade in telephones, and failed.

These able men found their plans thwarted by the indifference of the public, and often by open hostility. "The telephone is little better than a toy," said the Saturday Review; "it amazes ignorant people for a moment, but it is inferior to the well-established system of air-tubes." "What will become of the privacy of life?" asked another London editor. "What will become of the sanctity of the domestic hearth?" Writers vied with each other in inventing methods of pooh-poohing Bell and his invention. "It is ridiculously simple," said one. "It is only an electrical speaking-tube," said another. "It is a complicated form of speaking-trumpet," said a third. No British editor could at first conceive of any use for the telephone, except for divers and coal miners. The price, too, created a general outcry. Floods of toy telephones were being sold on the streets at a shilling apiece; and although the Government was charging sixty dollars a year for the use of its printing-telegraphs, people protested loudly against paying half as much for telephones. As late as 1882, Herbert Spencer writes: "The telephone is scarcely used at all in London, and is unknown in the other English cities."

The first man of consequence to befriend the telephone was Lord Kelvin, then an untitled young scientist. He had seen the original telephones at the Centennial in Philadelphia, and was so fascinated with them that the impulsive Bell had thrust them into his hands as a gift. At the next meeting of the British Association for the Advancement of Science, Lord Kelvin exhibited these. He did more. He became the champion of the telephone. He staked his reputation upon it. He told the story of the tests made at the Centennial, and assured the sceptical scientists that he had not been deceived. "All this my own ears heard," he said, "spoken to me with unmistakable distinctness by this circular disc of iron."

The scientists and electrical experts were, for the most part, split up into two camps. Some of them said the telephone was impossible, while others said that "nothing could be simpler." Almost all were agreed that what Bell had done was a humorous trifle. But Lord Kelvin persisted. He hammered the truth home that the telephone was "one of the most interesting inventions that has ever been made in the history of science." He gave a demonstration with one end of the wire in a coal mine. He stood side by side with Bell at a public meeting in Glasgow, and declared:

"The things that were called telephones before Bell were as different from Bell's telephone as a series of hand-claps are different from the human voice. They were in fact electrical claps; while Bell conceived the idea - THE WHOLLY ORIGINAL AND NOVEL IDEA - of giving continuity to the shocks, so as to perfectly reproduce the human voice."

One by one the scientists were forced to take the telephone seriously. At a public test there was one noted professor who still stood in the ranks of the doubters. He was asked to send a message. He went to the instrument with a grin of incredulity, and thinking the whole exhibition a joke, shouted into the mouthpiece: "Hi diddle diddle - follow up that." Then he listened for an answer. The look on his face changed to one of the utmost amazement. "It says - `The cat and the fiddle,'" he gasped, and forthwith he became a convert to telephony. By such tests the men of science were won over, and by the middle of 1877 Bell received a "vociferous welcome" when he addressed them at their annual convention at Plymouth.

Soon afterwards, The London Times surrendered. It whirled right-about-face and praised the telephone to the skies. "Suddenly and quietly the whole human race is brought within speaking and hearing distance," it exclaimed; "scarcely anything was more desired and more impossible." The next paper to quit the mob of scoffers was the Tatler, which said in an editorial peroration, "We cannot but feel impressed by the picture of a human child commanding the subtlest and strongest force in Nature to carry, like a slave, some whisper around the world."

Closely after the scientists and editors came the nobility. The Earl of Caithness led the way. He declared in public that "the telephone is the most extraordinary thing I ever saw in my life." And one wintry morning in 1878 Queen Victoria drove to the house of Sir Thomas Biddulph, in London, and for an hour talked and listened by telephone to Kate Field, who sat in a Downing Street office. Miss Field sang "Kathleen Mavourneen," and the Queen thanked her by telephone, saying she was "immensely pleased." She congratulated Bell himself, who was present, and asked if she might be permitted to buy the two telephones; whereupon Bell presented her with a pair done in ivory.

This incident, as may be imagined, did much to establish the reputation of telephony in Great Britain. A wire was at once strung to Windsor Castle. Others were ordered by the Daily News, the Persian Ambassador, and five or six lords and baronets. Then came an order which raised the hopes of the telephone men to the highest heaven, from the banking house of J. S. Morgan & Co. It was the first recognition from the "seats of the mighty" in the business and financial world. A tiny exchange, with ten wires, was promptly started in London; and on April 2d, 1879, Theodore Vail, the young manager of the Bell Company, sent an order to the factory in Boston, "Please make one hundred hand telephones for export trade as early as possible." The foreign trade had begun.

Then there came a thunderbolt out of a blue sky, a wholly unforeseen disaster. Just as a few energetic companies were sprouting up, the Postmaster General suddenly proclaimed that the telephone was a species of telegraph. According to a British law the telegraph was required to be a Government monopoly. This law had been passed six years before the telephone was born, but no matter. The telephone men protested and argued. Tyndall and Lord Kelvin warned the Government that it was making an indefensible mistake. But nothing could be done. Just as the first railways had been called toll-roads, so the telephone was solemnly declared to be a telegraph. Also, to add to the absurd humor of the situation, Judge Stephen, of the High Court of Justice, spoke the final word that compelled the telephone legally to be a telegraph, and sustained his opinion by a quotation from Webster's Dictionary, which was published twenty years before the telephone was invented.

Having captured this new rival, what next? The Postmaster General did not know. He had, of course, no experience in telephony, and neither had any of his officials in the telegraph department. There was no book and no college to instruct him. His telegraph was then, as it is to-day, a business failure. It was not earning its keep. Therefore he did not dare to shoulder the risk of constructing a second system of wires, and at last consented to give licenses to private companies.

But the muddle continued. In order to compel competition, according to the academic theories of the day, licenses were given to thirteen private companies. As might have been expected, the ablest company quickly swallowed the other twelve. If it had been let alone, this company might have given good service, but it was hobbled and fenced in by jealous regulations. It was compelled to pay one-tenth of its gross earnings to the Post Office. It was to hold itself ready to sell out at six months' notice. And as soon as it had strung a long-distance system of wires, the

Postmaster General pounced down upon it and took it away.

Then, in 1900, the Post Office tossed aside all obligations to the licensed company, and threw open the door to a free-for-all competition. It undertook to start a second system in London, and in two years discovered its blunder and proposed to cooperate. It granted licenses to five cities that demanded municipal ownership. These cities set out bravely, with loud beating of drums, plunged from one mishap to another, and finally quit. Even Glasgow, the premier city of municipal ownership, met its Waterloo in the telephone. It spent one million, eight hundred thousand dollars on a plant that was obsolete when it was new, ran it for a time at a loss, and then sold it to the Post Office in 1906 for one million, five hundred and twenty-five thousand dollars.

So, from first to last, the story of the telephone in Great Britain has been a "comedy of errors." There are now, in the two islands, not six hundred thousand telephones in use. London, with its six hundred and forty square miles of houses, has one-quarter of these, and is gaining at the rate of ten thousand a year. No large improvements are under way, as the Post Office has given notice that it will take over and operate all private companies on New Year's Day, 1912. The bureaucratic muddle, so it seems, is to continue indefinitely.

In Germany there has been the same burden of bureaucracy, but less backing and filling. There is a complete government monopoly. Whoever commits the crime of leasing telephone service to his neighbors may be sent to jail for six months. Here, too, the

Postmaster General has been supreme. He has forced the telephone business into a postal mould. The man in a small city must pay as high a rate for a small service, as the man in a large city pays for a large service. There is a fair degree of efficiency, but no high speed or record-breaking. The German engineers have not kept in close touch with the progress of telephony in the United States. They have preferred to devise methods of their own, and so have created a miscellaneous assortment of systems, good, bad, and indifferent. All told, there is probably an investment of seventy-five million dollars and a total of nine hundred thousand telephones.

Telephony has always been in high favor with the Kaiser. It is his custom, when planning a hunting party, to have a special wire strung to the forest headquarters, so that he can converse every morning with his Cabinet. He has conferred degrees and honors by telephone. Even his former Chancellor, Von Buelow, received his title of Count in this informal way. But the first friend of the telephone in Germany was Bismarck. The old Unifier saw instantly its value in holding a nation together, and ordered a line between his palace in Berlin and his farm at Varzin, which lay two hundred and thirty miles apart. This was as early as the Fall of 1877, and was thus the first long-distance line in Europe.

In France, as in England, the Government seized upon the telephone business as soon as the pioneer work had been done by private citizens. In 1889 it practically confiscated the Paris system, and after nine years of litigation paid five million francs to its owners. With this reckless beginning, it floundered from bad to worse. It assembled the most complete assortment of other nations' mistakes, and invented several of its own. Almost every known evil of bureaucracy was developed. The system of rates was turned upside down; the flat rate, which can be profitably permitted in small cities only, was put in force in the large cities, and the message rate, which is applicable only to large cities, was put in force in small places. The girl operators were entangled in a maze of civil service rules. They were not allowed to marry without the permission of the Postmaster General; and on no account might they dare to marry a mayor, a policeman, a cashier, or a foreigner, lest they betray the secrets of the switchboard.

There was no national plan, no standardization, no staff of inventors and improvers. Every user was required to buy his own telephone. As George Ade has said, "Anything attached to a wall is liable to be a telephone in Paris." And so, what with poor equipment and red tape, the French system became what it remains to-day, the most conspicuous example of what NOT to do in telephony.

There are barely as many telephones in the whole of France as ought normally to be in the city of Paris. There are not as many as are now in use in Chicago. The exasperated Parisians have protested. They have presented a petition with thirty-two thousand names. They have even organized a "Kickers' League" - the only body of its kind in any country - to demand good service at a fair price. The daily loss from bureaucratic telephony has become enormous. "One blundering girl in a telephone exchange cost me five thousand dollars on the day of the panic in 1907," said George Kessler. But the Government clears a net profit of three million dollars a year from its telephone monopoly; and until

1910, when a committee of betterment was appointed, it showed no concern at the discomfort of the public.

There was one striking lesson in telephone efficiency which Paris received in 1908, when its main exchange was totally destroyed by fire. "To build a new switchboard," said European manufacturers, "will require four or five months." A hustling young Chicagoan appeared on the scene. "We 'll put in a new switchboard in sixty days," he said; "and agree to forfeit six hundred dollars a day for delay." Such quick work had never been known. But it was Chicago's chance to show what she could do. Paris and Chicago are four thousand, five hundred miles apart, a twelve days' journey. The switchboard was to be a hundred and eighty feet in length, with ten thousand wires. Yet the Western Electric finished it in three weeks. It was rushed on six freight-cars to New York, loaded on the French steamer La Provence, and deposited at Paris in thirty-six days; so that by the time the sixty days had expired, it was running full speed with a staff of ninety operators.

Russia and Austria-Hungary have now about one hundred and twenty-five thousand telephones apiece. They are neck and neck in a race that has not at any time been a fast one. In each country the Government has been a neglectful stepmother to the telephone. It has starved the business with a lack of capital and used no enterprise in expanding it. Outside of Vienna, Budapest, St. Petersburg, and Moscow there are no wire-systems of any consequence. The political deadlock between Austria and Hungary shuts out any immediate hope of a happier life for the telephone in those countries; but in Russia there has recently been a change in policy that may open up a new era. Permits are now being offered to one private company in each city, in return for three per cent of the revenue. By this step Russia has unexpectedly swept to the front and is now, to telephone men, the freest country in Europe.

In tiny Switzerland there has been government ownership from the first, but with less detriment to the business than elsewhere. Here the officials have actually jilted the telegraph for the telephone. They have seen the value of the talking wire to hold their valley villages together; and so have cries-crossed the Alps with a cheap and somewhat flimsy system of telephony that carries sixty million conversations a year. Even the monks of St. Bernard, who rescue snowbound travellers, have now equipped their mountain with a series of telephone booths.

The highest telephone in the world is on the peak of Monte Rosa, in the Italian Alps, very nearly three miles above the level of the sea. It is linked to a line that runs to Rome, in order that a queen may talk to a professor. In this case the Queen is Margherita of Italy and the professor is Signor Mosso, the astronomer, who studies the heavens from an observatory on Monte Rosa. At her own expense, the Queen had this wire strung by a crew of linemen, who slipped and floundered on the mountain for six years before they had it pegged in place. The general situation in Italy is like that in Great Britain. The Government has always monopolized the long-distance lines, and is now about to buy out all private companies. There are only fifty-five thousand telephones to thirty-two million people - as many as in Norway and less than in Denmark. And in many of the southern and Sicilian provinces the jingle of the telephone bell is still an unfamiliar sound.

The main peculiarity in Holland is that there is no national plan, but rather a patchwork, that resembles Joseph's coat of many colors. Each city engineer has designed his own type of apparatus and had it made to order. Also, each company is fenced in by law within a six-mile circle, so that Holland is dotted with thumbnail systems, no two of which are alike. In Belgium there has been a government system since 1893, hence there is unity, but no enterprise. The plant is old-fashioned and too small. Spain has private companies, which give fairly good service to twenty thousand people. Roumania has half as many. Portugal has two small companies in Lisbon and Oporto. Greece, Servia, and Bulgaria have a scanty two thousand apiece. The frozen little isle of Iceland has one-quarter as many; and even into Turkey, which was a forbidden land under the regime of the old Sultan, the Young Turks are importing boxes of telephones and coils of copper wire.

There is one European country, and only one, which has caught the telephone spirit - Sweden. Here telephony had a free swinging start. It was let alone by the Post Office; and better still, it had a Man, a business-builder of remarkable force and ability, named Henry Cedergren. Had this man been made the Telephone-Master of Europe, there would have been a different story to tell. By his insistent enterprise he made Stockholm the best telephoned city outside of the United States. He pushed his country forward until, having one hundred and sixty-five thousand telephones, it stood fourth among the European nations. Since his death the Government has entered the field with a duplicate system, and a war has been begun which grows yearly more costly and absurd.

Asia, as yet, with her eight hundred and fifty million people, has fewer telephones than Philadelphia, and three-fourths of them are in the tiny island of Japan. The Japanese were enthusiastic telephonists from the first. They had a busy exchange in Tokio in 1883. This has now grown to have twenty-five thousand users, and might have more, if it had not been stunted by the peculiar policy of the Government. The public officials who operate the system are able men. They charge a fair price and make ten per cent profit for the State. But they do not keep pace with the demand. It is one of the oddest vagaries of public ownership that there is now in Tokio a WAITING LIST of eight thousand citizens, who are offering to pay for telephones and cannot get them. And when a Tokian dies, his franchise to a telephone, if he has one, is usually itemized in his will as a four-hundred-dollar property.

India, which is second on the Asiatic list, has no more than nine thousand telephones - one to every thirty-three thousand of her population! Not quite so many, in fact, as there are in five of the skyscrapers of New York. The Dutch East Indies and China have only seven thousand apiece, but in China there has recently come a forward movement. A fund of twenty million dollars is to be spent in constructing a national system of telephone and telegraph. Peking is now pointing with wonder and delight to a new exchange, spick and span, with a couple of ten-thousand-wire switchboards. Others are being built in Canton, Hankow, and Tien-Tsin. Ultimately, the telephone will flourish in China, as it has done in the Chinese quarter in San Francisco. The Empress of China, after the siege of Peking, commanded that a telephone should be hung in her palace, within reach of her dragon throne; and she was very friendly with any representative of the "Speaking

Lightning Sounds" business, as the Chinese term telephony.

In Persia the telephone made its entry recently in true comic-opera fashion. A new Shah, in an outburst of confidence, set up a wire between his palace and the market-place in Teheran, and invited his people to talk to him whenever they had grievances. And they talked! They talked so freely and used such language, that the Shah ordered out his soldiers and attacked them. He fired upon the new Parliament, and was at once chased out of Persia by the enraged people. From this it would appear that the telephone ought to be popular in Persia, although at present there are not more than twenty in use.

South America, outside of Buenos Ayres, has few telephones, probably not more than thirty thousand. Dom Pedro of Brazil, who befriended Bell at the Centennial, introduced telephony into his country in 1881; but it has not in thirty years been able to obtain ten thousand users. Canada has exactly the same number as Sweden - one hundred and sixty-five thousand. Mexico has perhaps ten thousand; New Zealand twenty-six thousand; and Australia fifty-five thousand.

Far down in the list of continents stands Africa. Egypt and Algeria have twelve thousand at the north; British South Africa has as many at the south; and in the vast stretches between there are barely a thousand more. Whoever pushes into Central Africa will still hear the beat of the wooden drum, which is the clattering sign-language of the natives. One strand of copper wire there is, through the Congo region, placed there by order of the late King of Belgium. To string it was probably the most adventurous piece of work in the history of telephone linemen. There was one seven hundred and fifty mile stretch of the central jungle. There were white ants that ate the wooden poles, and wild elephants that pulled up the iron poles. There were monkeys that played tag on the lines, and savages that stole the wire for arrow-heads. But the line was carried through, and to-day is alive with conversations concerning rubber and ivory.

So, we may almost say of the telephone that "there is no speech nor language where its voice is not heard." There are even a thousand miles of its wire in Abyssinia and one hundred and fifty miles in the Fiji Islands. Roughly speaking, there are now ten million telephones in all countries, employing two hundred and fifty thousand people, requiring twenty-one million miles of wire, representing a cost of fifteen hundred million dollars, and carrying fourteen thousand million conversations a year. All this, and yet the men who heard the first feeble cry of the infant telephone are still alive, and not by any means old.

No foreign country has reached the high American level of telephony. The United States has eight telephones per hundred of population, while no other country has one-half as many. Canada stands second, with almost four per hundred; and Sweden is third. Germany has as many telephones as the State of New York; and Great Britain as many as Ohio. Chicago has more than London; and Boston twice as many as Paris. In the whole of Europe, with her twenty nations, there are one-third as many telephones as in the United States. In proportion to her population, Europe has only one-thirteenth as many.

The United States writes half as many letters as Europe, sends one-third as many telegrams, and talks twice as much at the telephone. The average European family sends three telegrams a year, and three letters and one telephone message a week; while the average American family sends five telegrams a year, and seven letters and eleven telephone messages a week. This one nation, which owns six per cent of the earth and is five per cent of the human race, has SEVENTY per cent of the telephones. And fifty per cent, or one-half, of the telephony of the world, is now comprised in the Bell System of this country.

There are only six nations in Europe that make a fair showing - the Germans, British, Swedish, Danes, Norwegians, and Swiss. The others have less than one telephone per hundred. Little Denmark has more than Austria. Little Finland has better service than France. The Belgian telephones have cost the most - two hundred and seventy-three dollars apiece; and the Finnish telephones the least - eighty-one dollars. But a telephone in Belgium earns three times as much as one in Norway. In general, the lesson in Europe is this, that the telephone is what a nation makes it. Its usefulness depends upon the sense and enterprise with which it is handled. It may be either an invaluable asset or a nuisance.

Too much government! That has been the basic reason for failure in most countries. Before the telephone was invented, the telegraph had been made a State monopoly; and the telephone was regarded as a species of telegraph. The public officials did not see that a telephone system is a highly complex and technical problem, much more like a piano factory or a steel-mill. And so, wherever a group of citizens established a telephone service, the government officials looked upon it with jealous eyes, and usually snatched it away. The telephone thus became a part of the telegraph, which is a part of the post office, which is a part of the government. It is a fraction of a fraction of a fraction - a mere twig of bureaucracy. Under such conditions the telephone could not prosper. The wonder is that it survived.

Handled on the American plan, the telephone abroad may be raised to American levels. There is no racial reason for failure. The slow service and the bungling are the natural results of treating the telephone as though it were a road or a fire department; and any nation that rises to a proper conception of the telephone, that dares to put it into competent hands and to strengthen it with enough capital, can secure as alert and brisk a service as heart can wish. Some nations are already on the way. China, Japan, and France have sent delegations to New York City - "the Mecca of telephone men," to learn the art of telephony in its highest development. Even Russia has rescued the telephone from her bureaucrats and is now offering it freely to men of enterprise.

In most foreign countries telephone service is being steadily geared up to a faster pace. The craze for "cheap and nasty" telephony is passing; and the idea that the telephone is above all else a SPEED instrument, is gaining ground. A faster long-distance service, at double rates, is being well patronized. Slow-moving races are learning the value of time, which is the first lesson in telephony. Our reapers and mowers now go to seventy-five nations. Our street cars run in all great cities. Morocco is importing our dollar watches; Korea is learning the waste of allowing nine men to dig with one spade. And all this means telephones.

In thirty years, the Western Electric has sold sixty-seven million dollars' worth of telephonic apparatus to foreign countries. But this is no more than a fair beginning. To put one telephone in China to every hundred people will mean an outlay of three hundred million dollars. To give Europe as fit an equipment as the United States now has, will mean thirty million telephones, with proper wire and switchboards to match. And while telephony for the masses is not yet a live question in many countries, sooner or later, in the relentless push of civilization, it must come.

Possibly, in that far future of peace and goodwill among nations, when each country does for all the others what it can do best, the United States may be generally recognized as the source of skill and authority on telephony. It may be called in to rebuild or operate the telephone systems of other countries, in the same way that it is now supplying oil and steel rails and farm machinery. Just as the wise buyer of to-day asks France for champagne, Germany for toys, England for cottons, and the Orient for rugs, so he will learn to look upon the United States as the natural home and headquarters of the telephone.

CHAPTER IX

THE FUTURE OF THE TELEPHONE

In the Spring of 1907 Theodore N. Vail, a rugged, ruddy, white-haired man, was superintending the building of a big barn in northern Vermont. His house stood near-by, on a balcony of rolling land that overlooked the town of Lyndon and far beyond, across evergreen forests to the massive bulk of Burke Mountain. His farm, very nearly ten square miles in area, lay back of the house in a great oval of field and woodland, with several dozen cottages in the clearings. His Welsh ponies and Swiss cattle were grazing on the May grass, and the men were busy with the ploughs and harrows and seeders. It was almost thirty years since he had been called in to create the business structure of telephony, and to shape the general plan of its development. Since then he had done many other things. The one city of Buenos Ayres had paid him more, merely for giving it a system of trolleys and electric lights, than the United States had paid him for putting the telephone on a business basis. He was now rich and retired, free to enjoy his play-work of the farm and to forget the troubles of the city and the telephone.

But, as he stood among his barn-builders, there arrived from Boston and New York a delegation of telephone directors. Most of them belonged to the "Old Guard" of telephony. They had fought under Vail in the pioneer days; and now they had come to ask him to return to the telephone business, after twenty years of absence. Vail laughed at the suggestion.

"Nonsense," he said, "I'm too old. I'm sixty-two years of age." The directors persisted. They spoke of the approaching storm-cloud of panic and the need of another strong hand at the wheel until the crisis was over, but Vail still refused. They spoke of old times and old memories, but he shook his head. "All my life," he said, "I have wanted to be a farmer."

Then they drew a picture of the telephone situation. They showed him that the "grand telephonic system" which he had planned was unfinished. He was its architect, and it was undone. The telephone business was energetic and prosperous. Under the brilliant leadership of Frederick P. Fish, it had grown by leaps and bounds. But it was still far from being the SYSTEM that Vail had dreamed of in his younger days; and so, when the directors put before him his unfinished plan, he surrendered. The instinct for completeness, which is one of the dominating characteristics of his mind, compelled him to consent. It was the call of the telephone.

Since that May morning, 1907, great things have been done by the men of the telephone and telegraph world. The Bell System was brought through the panic without a scratch. When the doubt and confusion were at their worst, Vail wrote an open letter to his stockholders, in his practical, farmer-like way. He said:

"Our net earnings for the last ten months were $13,715,000, as against $11,579,000 for the same period in 1906. We have now in the banks over $18,000,000; and we will not need to borrow any money for two years."

Soon afterwards, the work of consolidation began. Companies that overlapped were united. Small local wire-clusters, several thousands of them, were linked to the national lines. A policy of publicity superseded the secrecy which had naturally grown to be a habit in the days of patent litigation. Visitors and reporters found an open door. Educational advertisements were published in the most popular magazines. The corps of inventors was spurred up to conquer the long-distance problems. And in return for a thirty million check, the control of the historic Western Union was transferred from the children of Jay Gould to the thirty thousand stock-holders of the American Telephone and Telegraph Company.

From what has been done, therefore, we may venture a guess as to the future of the telephone. This "grand telephonic system" which had no existence thirty years ago, except in the imagination of Vail, seems to be at hand. The very newsboys in the streets are crying it. And while there is, of course, no exact blueprint of a best possible telephone system, we can now see the general outlines of Vail's plan.

There is nothing mysterious or ominous in this plan. It has nothing to do with the pools and conspiracies of Wall Street. No one will be squeezed out except the promoters of paper companies. The simple fact is that Vail is organizing a complete Bell System for the same reason that he built one big comfortable barn for his Swiss cattle and his Welsh ponies, instead of half a dozen small uncomfortable sheds. He has never been a "high financier" to juggle profits out of other men's losses. He is merely applying to the telephone business the same hard sense that any farmer uses in the management of his farm. He is building a Big Barn, metaphorically, for the telephone and telegraph.

Plainly, the telephone system of the future will be national, so that any two people in the same country will be able to talk to one another. It will not be competitive, for the reason that no farmer would think for a moment of running his farm on competitive lines. It will have a staff-and-line organization, to use a military phrase. Each local company will continue to handle its own local affairs, and exercise to the full the basic virtue of self-help. But there will also be, as now, a central body of experts to handle the larger affairs that are common to all companies. No separateness or secession on the one side, nor bureaucracy on the other - that is the typically American idea that underlies the ideal telephone system.

The line of authority, in such a system, will begin with the local manager. From him it will rise to the directors of the State company; then higher still to the directors of the national company; and finally, above all corporate leaders to the Federal Government itself. The failure of government ownership of the telephone in so many foreign countries does not mean that the private companies will have absolute power. Quite the reverse. The lesson of thirty years' experience shows that a private telephone company is apt to be much more obedient to the will of the people than if it were a Government department. But it is an axiom of democracy that no company, however well conducted, will be permitted to control a public convenience without being held strictly responsible for its own acts.

As politics becomes less of a game and more of a responsibility, the telephone of the future will doubtless be supervised by some sort of public committee, which will have power to pass upon complaints, and to prevent the nuisance of duplication and the swindle of watering stock.

As this Federal supervision becomes more and more efficient, the present fear of monopoly will decrease, just as it did in the case of the railways. It is a fact, although now generally forgotten, that the first railways of the United States were run for ten years or more on an anti-monopoly plan. The tracks were free to all. Any one who owned a cart with flanged wheels could drive it on the rails and compete with the locomotives. There was a happy-go-lucky jumble of trains and wagons, all held back by the slowest team; and this continued on some railways until as late as 1857. By that time the people saw that competition on a railway track was absurd. They allowed each track to be monopolized by one company, and the era of expansion began.

No one, certainly, at the present time, regrets the passing of the independent teamster. He was much more arbitrary and expensive than any railroad has ever dared to be; and as the country grew, he became impossible. He was not the fittest to survive. For the general good, he was held back from competing with the railroad, and taught to cooperate with it by hauling freight to and from the depots. This, to his surprise, he found much more profitable and pleasant. He had been squeezed out of a bad job into a good one. And by a similar process of evolution, the United States is rapidly outgrowing the small independent telephone companies. These will eventually, one by one, rise as the teamster did to a higher social value, by clasping wires with the main system of telephony.

Until 1881 the Bell System was in the hands of a family group. It was a strictly private enterprise. The public had been asked to help in its launching, and had refused. But after 1881 it passed into the control of the small stock-holders, and has remained there without a break. It is now one of our most democratized businesses, scattering either wages or dividends into more than a hundred thousand homes. It has at times been exclusive, but never sordid. It has never been dollar-mad, nor frenzied by the virus of stock-gambling. There has always been a vein of sentiment in it that kept it in touch with human nature. Even at the present time, each check of the American Telephone and Telegraph Company carries on it a picture of a pretty Cupid, sitting on a chair upon which he has placed a thick book, and gayly prattling into a telephone.

Several sweeping changes may be expected in the near future, now that there is team-play between the Bell System and the Western Union. Already, by a stroke of the pen, five million users of telephones have been put on the credit books of the Western Union; and every Bell telephone office is now a telegraph office. Three telephone messages and eight telegrams may be sent AT THE SAME TIME over two pairs of wires: that is one of the recent miracles of science, and is now to be tried out upon a gigantic scale. Most of the long-distance telephone wires, fully two million miles, can be used for telegraphic purposes; and a third of the Western Union wires, five hundred thousand miles, may with a few changes be used for talking.

The Western Union is paying rent for twenty-two thousand, five hundred offices, all of which helps to make telegraphy a luxury of the few. It is employing as large a force of messenger-boys as the army that marched with General Sherman from Atlanta to the sea. Both of these items of expense will dwindle when a Bell wire and a Morse wire can be brought to a common terminal; and when a telegram can be received or delivered by telephone. There will also be a gain, perhaps the largest of all, in removing the trudging little messenger-boy from the streets and sending him either to school or to learn some useful trade.

The fact is that the United States is the first country that has succeeded in putting both telephone and telegraph upon the proper basis.

Elsewhere either the two are widely apart, or the telephone is a mere adjunct of a telegraphic department. According to the new American plan, the two are not competitive, but complementary. The one is a supplement to the other. The post office sends a package; the telegraph sends the contents of the package; but the telephone sends nothing. It is an apparatus that makes conversation possible between two separated people. Each of the three has a distinct field of its own, so that there has never been any cause for jealousy among them.

To make the telephone an annex of the post office or the telegraph has become absurd. There are now in the whole world very nearly as many messages sent by telephone as by letter; and there are THIRT-TWO TIMES as many telephone calls as telegrams. In the United States, the telephone has grown to be the big brother of the telegraph. It has six times the net earnings and eight times the wire. And it transmits as many messages as the combined total of telegrams, letters, and railroad passengers.

This universal trend toward consolidation has introduced a variety of problems that will engage the ablest brains in the telephone world for many years to come. How to get the benefits of organization without its losses, to become strong without losing quickness, to become systematic without losing the dash and dare of earlier days, to develop the working force into an army of high-speed specialists without losing the bird's-eye view of the whole situation, - these are the riddles of the new type, for which the telephonists of the next generation must find the answers. They illustrate the nature of the big jobs that the telephone has to offer to an ambitious and gifted young man of to-day.

"The problems never were as large or as complex as they are right now," says J. J. Carty, the chief of the telephone engineers. The eternal struggle remains between the large and little ideas - between the men who see what might be and the men who only see what IS. There is still the race to break records. Already the girl at the switchboard can find the person wanted in thirty seconds. This is one-tenth of the time that was taken in the early centrals; but it is still too long. It is one-half of a valuable minute. It must be cut to twenty-five seconds, or twenty or fifteen.

There is still the inventors' battle to gain miles. The distance over which conversations can be held has been increased from twenty miles to twenty-five hundred. But this is not far enough. There are some civilized human beings who are twelve thousand miles apart, and who have interests in common. During the Boxer Rebellion in China, for instance, there were Americans in Peking who would gladly have given half of their fortune for the use of a pair of wires to New York.

In the earliest days of the telephone, Bell was fond of prophesying that "the time will come when we will talk across the Atlantic Ocean"; but this was regarded as a poetical fancy until Pupin invented his method of automatically propelling the electric current. Since then the most conservative engineer will discuss the problem of transatlantic telephony. And as for the poets, they are now dreaming of the time when a man may speak and hear his own voice come back to him around the world.

The immediate long-distance problem is, of course, to talk from New York to the Pacific. The two oceans are now only three and a half days apart by rail. Seattle is clamoring for a wire to the East. San Diego wants one in time for her Panama Canal Exposition in 1915. The wires are already strung to San Francisco, but cannot be used in the present stage of the art. And Vail's captains are working now with almost breathless haste to give him a birthday present of a talk across the continent from his farm in Vermont.

"I can see a universal system of telephony for the United States in the very near future," says Carty. "There is a statue of Seward standing in one of the streets of Seattle. The inscription upon it is, 'To a United Country.' But as an Easterner stands there, he feels the isolation of that Far Western State, and he will always feel it, until he can talk from one side of the United States to the other. For my part," continues Carty, "I believe we will talk across continents and across oceans. Why not? Are there not more cells in one human body than there are people in the whole earth?"

Some future Carty may solve the abandoned problem of the single wire, and cut the copper bill in two by restoring the grounded circuit. He may transmit vision as well as speech. He may perfect a third-rail system for use on moving trains. He may conceive of an ideal insulating material to supersede glass, mica, paper, and enamel. He may establish a universal code, so that all persons of importance in the United States shall have call-numbers by which they may instantly be located, as books are in a library.

Some other young man may create a commercial department on wide lines, a work which telephone men have as yet been too specialized to do. Whoever does this will be a man of comprehensive brain. He will be as closely in touch with the average man as with the art of telephony. He will know the gossip of the street, the demands of the labor unions, and the policies of governors and presidents. The psychology of the Western farmer will concern him, and the tone of the daily press, and the methods of department stores. It will be his aim to know the subtle chemistry of public opinion, and to adapt the telephone service to the shifting moods and necessities of the times. HE WILL FIT TELEPHONY LIKE A GARMENT AROUND THE HABITS OF THE PEOPLE.

Also, now that the telephone business has become strong, its next anxiety must be to develop the virtues, and not the defects, of strength. Its motto must be "Ich dien" - I serve; and it will be the work of the future statesmen of the telephone to illustrate this motto in all its practical variations. They will cater and explain, and explain and cater. They will educate and educate, until they have created an expert public. They will teach by pictures and lectures and exhibitions. They will have charts and diagrams hung in the telephone booths, so that the person who is waiting for a call may learn a little and pass the time more pleasantly. They will, in a word, attend to those innumerable trifles that make the perfection of public service.

Already the Bell System has gone far in this direction by organizing what might fairly be called a foresight department. Here is where the fortune-tellers of the business sit. When new lines or exchanges are to be built, these men study the situation with an eye to the future. They prepare a "fundamental plan," outlining what may reasonably be expected to happen in fifteen or twenty years. Invariably they are optimists. They make provision for growth, but none at all for shrinkage. By their advice, there is now twenty-five million dollars' worth of reserve plant in the various Bell Companies, waiting for the country to grow up to it. Even in the city of New York, one-half of the cable ducts are empty, in expectation of the greater city of eight million population which is scheduled to arrive in 1928. There are perhaps few more impressive evidences of practical optimism and confidence than a new telephone exchange, with two-thirds of its wires waiting for the business of the future.

Eventually, this foresight department will expand. It may, if a leader of genius appear, become the first real corps of practical sociologists, which will substitute facts for the present hotch-potch of theories. It will prepare a "fundamental plan" of the whole United States, showing the centre of each industry and the main runways of traffic. It will act upon the basic fact that WHEREVER THERE IS INTERDEPENDENCE, THERE IS BOUND TO BE TELEPHONY; and it will therefore prepare maps of interdependence, showing the widely scattered groups of industry and finance, and the lines that weave them into a pattern of national cooperation.

As yet, no nation, not even our own, has seen the full value of the long-distance telephone. Few have the imagination to see what has been made possible, and to realize that an actual face-to-face conversation may take place, even though there be a thousand miles between. Neither can it seem credible that a man in a distant city may be located as readily as though he were close at hand. It is too amazing to be true, and possibly a new generation will have to arrive before it will be taken for granted and acted upon freely. Ultimately, there can be no doubt that long-distance telephony will be regarded as a national asset of the highest value, for the reason that it can prevent so much of the enormous economic waste of travel.

Nothing that science can say will ever decrease the marvel of a long-distance conversation, and there may come in the future an Interpreter who will put it before our eyes in the form of a moving-picture. He will enable us to follow the flying words in a talk from Boston to Denver. We will flash first to Worcester, cross the Hudson on the high bridge at Poughkeepsie, swing southwest through a dozen coal towns to the outskirts of Philadelphia, leap across the Susquehanna, zigzag up and down the Alleghenies into the murk of Pittsburg, cross the Ohio at Wheeling, glance past

Columbus and Indianapolis, over the Wabash at Terre Haute, into St. Louis by the Eads bridge, through Kansas City, across the Missouri, along the corn-fields of Kansas, and then on - on - on with the Sante Fe Railway, across vast plains and past the brink of the Grand Canyon, to Pueblo and the lofty city of Denver. Twenty-five hundred miles along a thousand tons of copper wire! From Bunker Hill to Pike's Peak IN A SECOND!

Herbert Spencer, in his autobiography, alludes to the impressive fact that while the eye is reading a single line of type, the earth has travelled thirty miles through space. But this, in telephony, would be slow travelling. It is simple everyday truth to say that while your eye is reading this dash, - , a telephone sound can be carried from New York to Chicago.

There are many reasons to believe that for the practical idealists of the future, the supreme study will be the force that makes such miracles possible. Six thousand million dollars - one-twentieth of our national wealth - is at the present time invested in electrical development. The Electrical Age has not yet arrived; but it is at hand; and no one can tell how brilliant the result may be, when the creative minds of a nation are focussed upon the subdual of this mysterious force, which has more power and more delicacy than any other force that man has been able to harness.

As a tame and tractable energy, Electricity is new. It has no past and no pedigree. It is younger than many people who are now alive. Among the wise men of Greece and Rome, few knew its existence, and none put it to any practical use. The wisest knew that a piece of amber, when rubbed, will attract feathery substances. But they regarded this as poetry rather than science. There was a pretty legend among the Phoenicians that the pieces of amber were the petrified tears of maidens who had thrown themselves into the sea because of unrequited love, and each bead of amber was highly prized. It was worn as an amulet and a symbol of purity. Not for two thousand years did any one dream that within its golden heart lay hidden the secret of a new electrical civilization.

Not even in 1752, when Benjamin Franklin flew his famous kite on the banks of the Schuylkill River, and captured the first CANNED LIGHTNING, was there any definite knowledge of electrical energy. His lightning-rod was regarded as an insult to the deity of Heaven. It was blamed for the earthquake of 1755. And not until the telegraph of Morse came into general use, did men dare to think of the thunder-bolt of Jove as a possible servant of the human race.

Thus it happened that when Bell invented the telephone, he surprised the world with a new idea. He had to make the thought as well as the thing. No Jules Verne or H. G. Wells had foreseen it. The author of the Arabian Nights fantasies had conceived of a flying carpet, but neither he nor any one else had conceived of flying conversation. In all the literature of ancient days, there is not a line that will apply to the telephone, except possibly that expressive phrase in the Bible, "And there came a voice." In these more privileged days, the telephone has come to be regarded as a commonplace fact of everyday life; and we are apt to forget that the wonder of it has become greater and not less; and that there are still honor and profit, plenty of both, to be won by the inventor and the scientist.

The flood of electrical patents was never higher than now. There are literally more in a single month than the total number issued by the Patent Office up to 1859. The Bell System has three hundred experts who are paid to do nothing else but try out all new ideas and inventions; and before these words can pass into the printed book, new uses and new methods will have been discovered. There is therefore no immediate danger that the art of telephony will be less fascinating in the future than it has been in the past. It will still be the most alluring and elusive sprite that ever led the way through a Dark Continent of mysterious phenomena.

There still remains for some future scientist the task of showing us in detail exactly what the telephone current does. Such a man will study vibrations as Darwin studied the differentiation of species. He will investigate how a child's voice, speaking from Boston to Omaha, can vibrate more than a million pounds of copper wire; and he will invent a finer system of time to fit the telephone, which can do as many different things in a second as a man can do in a day, transmitting with every tick of the clock from twenty-five to eighty thousand vibrations. He will deal with the various vibrations of nerves and wires and wireless air, that are necessary in conveying thought between two separated minds. He will make clear how a thought, originating in the brain, passes along the nerve-wires to the vocal chords, and then in wireless vibration of air to the disc of the transmitter. At the other end of the line the second disc re-creates these vibrations, which impinge upon the nerve-wires of an ear, and are thus carried to the consciousness of another brain.

And so, notwithstanding all that has been done since Bell opened up the way, the telephone remains the acme of electrical marvels. No other thing does so much with so little energy. No other thing is more enswathed in the unknown. Not even the gray-haired pioneers who have lived with the telephone since its birth, can understand their protege. As to the why and the how, there is as yet no answer. It is as true of telephony to-day as it was in 1876, that a child can use what the wisest sages cannot comprehend.

Here is a tiny disc of sheet-iron. I speak - it shudders. It has a different shudder for every sound. It has thousands of millions of different shudders. There is a second disc many miles away, perhaps twenty-five hundred miles away. Between the two discs runs a copper wire. As I speak, a thrill of electricity flits along the wire. This thrill is moulded by the shudder of the disc. It makes the second disc shudder. And the shudder of the second disc reproduces my voice. That is what happens. But how - not all the scientists of the world can tell.

The telephone current is a phenomenon of the ether, say the theorists. But what is ether? No one knows. Sir Oliver Lodge has guessed that it is "perhaps the only substantial thing in the material universe"; but no one knows. There is nothing to guide us in that unknown country except a sign-post that points upwards and bears the one word - "Perhaps." The ether of space! Here is an Eldorado for the scientists of the future, and whoever can first map it out will go far toward discovering the secret of telephony.

Some day - who knows? - there may come the poetry and grand opera of the telephone. Artists may come who will portray the marvel of the wires that quiver with electrified words, and the romance of the switchboards that tremble with the secrets of a great city. Already Puvis de Chavannes, by one of his superb panels in the Boston Library, has admitted the telephone and the telegraph to the world of art. He has embodied them as two flying figures, poised above the electric wires, and with the following inscription underneath: "By the wondrous agency of electricity, speech dashes through space and swift as lightning bears tidings of good and evil."

But these random guesses as to the future of the telephone may fall far short of what the reality will be. In these dazzling days it is idle to predict. The inventor has everywhere put the prophet out of business. Fact has outrun Fancy. When Morse, for instance, was tacking up his first little line of wire around the Speedwell Iron Works, who could have foreseen two hundred and fifty thousand miles of submarine cables, by which the very oceans are all aquiver with the news of the world? When Fulton's tiny tea-kettle of a boat steamed up the Hudson to Albany in two days, who could have foreseen the steel leviathans, one-sixth of a mile in length, that can in the same time cut the Atlantic Ocean in halves? And when Bell stood in a dingy workshop in Boston and heard the clang of a clock-spring come over an electric wire, who could have foreseen the massive structure of the Bell System, built up by half the telephones of the world, and by the investment of more actual capital than has gone to the making of any other industrial association? Who could have foreseen what the telephone bells have done to ring out the old ways and to ring in the new; to ring out delay, and isolation and to ring in the efficiency and the friendliness of a truly united people?

Printed in the United States
41697LVS00004B/227